DATE DUE

1-12-81			
MAR 9 1981			
MAR 3 0 1981			
APR 2 0 1981			
APR 2 7 1981			
SEP 0 7 1999			

GAYLORD 234   PRINTED IN U. S. A.

81   O

MANUAL OF COLOUR REPRODUCTION
FOR PRINTING AND THE GRAPHIC ARTS

Translation: Edward S. Bomback

Thanks are due to Mr. Rino Ferrotti, Mr. Iginio Astori and in particular to Signorina Carla Morazzo for their collaboration. All the drawings, graphs and colour separations used in this book have been produced at the School of Graphic Arts MIMEP of Pessano who hold the copyright.

Massimo Astrua

MANUAL OF COLOUR REPRODUCTION
for Printing and the Graphic Arts

FOUNTAIN PRESS

Fountain Press
Model and Allied Publications Ltd.,
Book Division,
Station Road, Kings Langley,
Hertfordshire, England

First Published in England, 1973

© 1973 Massimo Astrua, Pessano, Italy

ISBN 0 85242 004 8

Printed in Italy by Scoula Grafica Ist., San Gaetano, Vicenza

PREFACE

In compiling this book we have sought to offer a clear and as far as possible complete synthesis of the theorectical principles which govern and explain the practice of photomechanical colour reproduction.

The book is intended for readers who have received a secondary education, but even those with less acquaintance of elementary mathematics and science should, with a little effort, be able to understand and benefit from a study of the book. To this end we have made ample use of diagrams and illustrations throughout the text and have also inserted simplified explanations or footnotes of every new idea before applying it to the subject matter of the book.

We trust that our efforts — which reflect the courses given to the students at the School of Graphic Arts "MIMEP" — may also prove helpful to other students and young workers in the photomechanical arts who wish to improve their knowledge and technique.

Pessano, 1972

Don Massimo Astrua

CONTENTS

SENSITOMETRY

PART I

SENSITOMETRY

Sensitometry (= measurement of sensitivity) is the science that studies and measures the effect of light on a photographic emulsion.

The essential product of these effects is the *blackening* which the emulsion acquires during development in the areas which have been exposed to light.

This blackening will be "denser" in the areas which have received a greater amount of light and less "dense" in those which have received less light.

In order to measure the sensitivity to light of a photographic emulsion (which is, in fact, the object of sensitometry), we must firstly be familiar with the photographic emulsion; we must be able to measure the light which has caused its blackening, and finally, we must know how to measure this same blackening, that is, its density.

These three requirements form the basis of the chapters in this first part of the book:

1st The photographic emulsion
2nd The measurement of the light (Photometry)
3rd The measurement of the density (Densitometry).

CHAPTER 1

THE PHOTOGRAPHIC EMULSION

It is impossible to understand much that we shall have to say about the photographic emulsion and about its chemical behaviour without having some idea of elementary chemistry. We invite even those who are already familiar with it to study this brief summary, since the examples given are of special interest in the chemistry of photography and thus have a bearing on the subject matter of this book.

INSERTION I

IDEAS OF BASIC CHEMISTRY

PARTICLES - ATOMS - MOLECULES

a) All existing matter is an enormous agglomeration of very small particles of matter, so variously combined among themselves as to give place to all the different substances existing in nature.

These particles are of three kinds: protons, neutrons and electrons.

b) They are disposed in a similar way to the planets of the solar system; at the centre the protons and neutrons, firmly united among themselves, take the place of the sun and form the nucleus: in orbit around them turn the electrons like microscopic planets.

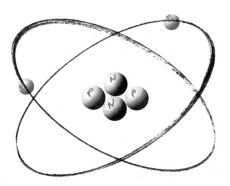

Fig. 1

This minute solar system is called an ATOM.

At the present time we know about a hundred atoms, one for each of the different elements existing in nature, such as the atom of iron, the atom of oxygen, the atom of silver and so on.

12

c) When two or more atoms combine, they give rise to a new form of matter called a MOLECULE. For example, two atoms of hydrogen (symbol H) and one atom of oxygen (symbol O) combine to form the molecule of water (symbol H_2O).

A vast number of molecules are known, and there is one for every substance existing in nature.

Let us now examine in detail the atom and the molecule so as to obtain some basic ideas which will help us to understand something of the chemistry of photography.

THE ATOM

a) Structure of the atom

As we have already said, the atom is similar to a minute solar system: at the centre there is a nucleus around which the electrons rotate. At the same time, the nucleus is made up of protons to which are normally united the neutrons.

Let us take for an example the atom of helium so as to obtain some idea, even if imperfect, of how the atom is made up:

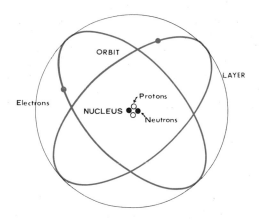

Fig. 2

The nucleus is composed of 2 protons and 2 neutrons.

Around the nucleus, and at the same distance from it, rotate 2 electrons forming the first "orbit" (or first electronic layer).

These layers can be as many as 7: the first cannot contain more than 2 electrons, the second not more than 8, the third not more than 18,... then things become more complex. It is enough for us to know that the outermost layer, whichever it be, cannot exceed the number of 8 electrons.

The same atom of helium can be represented schematically so:

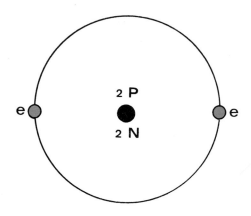

Fig. 3

To have some idea of the smallness of the atom we could imagine the nucleus as having a diameter of one centimetre and if it were placed at the centre of the Piazza San Pietro in Rome, the electrons would be invisible and would be rotating in the region of the columns. Physicists have compared the ratio of the nuclear matter and the orbital space with that of the sun and the space that separates it from the orbits of the planets. Matter in thus very much a system of emptiness: if we could squash the Earth so as to press one nucleus to the next, it would become no larger than an orange while at the same time retaining its enormous mass or weight!

b) Atomic particles

The principal, basic particles which form an atom are therefore three:

1) *The protons*, with mass (weight) relatively great and with a fixed positive ($\oplus$) charge of electricity.

2) *The neutrons*, having the same mass (weight) as the protons, but without a charge of electricity (neutral).

3) *The electrons*, of negligible mass (weight) — only $\frac{1}{1846}$ of the proton — and with an electric charge equal to that of the protons, but negative ($\ominus$).

14

c) The number of atomic particles

1) That which characterises an element (that is, what makes an element what it is) is *the number of its protons*. Every element has its "own" characteristic number of protons, from hydrogen which has 1, to lawrencium which has 103. Thus the atoms of all the known elements (about a hundred) can be listed according to their number of protons — or "atomic number".

2) *The number of neutrons* of the nucleus is not always equal to the number of protons and is thus not a characteristic of the element. The atom of an element can, in fact, have a different number of neutrons while still remaining that element (such an atom takes the name "isotope") of the nucleus, but there is a variety of carbon that has 7 neutrons in the nucleus and is called isotope 7 of carbon.

3) *The number of electrons* which rotate round the nucleus in an electrically neutral atom is always equal to the number of protons: in order for an atom to be neutral, the number of electrons $\ominus$ must be the same as the number of protons $\oplus$. (See footnote 1). For example, hydrogen which has only one proton will have one electron; helium will have 2, and lawrencium 103.

Sometimes atoms lose or gain (for reasons which we shall see) one or more electrons in the last layer: in this case the atom does not change its own nature (in fact the number of protons is unchanged), but loses its electrical neutrality, becoming positive (if it loses electrons) or negative (if it gains electrons). These variations of the same element are called "ions" of that element. Thus the atom of silver Ag (neutral), which has 47 protons in the nucleus and 47 electrons in the orbits, will still remain silver (47 protons) if it loses one electron but will have one negative electric charge less (46 electrons) and will thus become a positive ion of silver (Ag^+).

The arrangement of the electrons round the nucleus is very important and particularly the number of electrons in the outermost orbit, since on this (as we shall see) depends to a large extent the physical and chemical properties of the element.

Footnote (1) Let us recall the relations of forces between electrically charged particles:
a) Electric charges with different signs attract each other.

$$\oplus \rightarrow \qquad \leftarrow \ominus$$

b) Electric charges with the same sign repell each other.

$$\leftarrow \oplus \qquad \oplus \rightarrow$$
$$\leftarrow \ominus \qquad \ominus \rightarrow$$

c) The basic positive electric charge (that possessed by the proton) and the basic negative electric charge (that possessed by the electron) neutralise each other in turn.

SUMMARY

		Electric charge	Mass (weight)	Number
ATOM	NUCLEUS → PROTONS	positive electric charge ⊕	heavy	Their number is characteristic (atomic number)
	NUCLEUS → NEUTRONS	neutral Ⓝ	same as protons	Their number added to that of the protons gives the weight of the element (Atomic weight) (if the number is different we have *isotopes*)
	ORBITS (max. 7) → ELECTRONS	negative electric charge ⊖	very light ($\frac{1}{1846}$ of the protons)	Their *total* number (in the neutral atom) is equal to that of the protons; (if the number is different we get *ions*) Their number in *the outermost orbit* determines the physical and chemical properties of the element

Fig. 4

16

THE PERIODIC SYSTEM OF THE ELEMENTS

This is the complete and rational arrangement of all the elements (atoms) according to their characteristics, devised by Mendeleev and later perfected by other scientists (see fig. 5).

It is useful to examine briefly six aspects of it:

1) By arranging the elements from left to right according to their atomic number (1, 2, 3, 4...) and placing them in horizontal lines according the number of electronic layers, we find at the end of each line (or period), forming a column one above the other (groups), all the "noble gases", that is those elements which, having 8 electrons on the outermost layer, are *saturated* and chemically inert. (Helium, having only a single layer, is saturated with 2 electrons).

2) In this arrangement, the elements in the 18 vertical columns (or groups) show the same physical and chemical characteristics, having in the outermost layer (save for certain exceptions) the same number of electrons.
These groups are labelled by Roman numbers (from I to VIII) followed with a capital letter (A or B) to indicate the corresponding sub-group. Thus the elements of group VII A (called the Halogens) behave in the same way chemically.

3) The elements arranged on the same horizontal line (or period) do not possess the same physical and chemical characteristics, but have the same number of electronic layers and tend to shape their outermost layer to that of the noble gas which closes the period.
On this *tendency* to copy the noble gases we shall return later when we examine the way by which atoms combine to form molecules (chemical links and valency).

4) The elements may be divided into two large categories: those called METALS having the three characteristics of *compactness, lustre and conductibility* of heat and electricity (these are in the great majority, more than two-thirds, and are shown in yellow in fig. 5); and those called NON-METALS which have the opposite characteristics of metals.
Note, however, that there is no clear-cut division between metals and non-metals but rather a gradation of elements with mixed characteristics, called Metalloids.
In the periodic table the most metallic elements are situated low down to the left, those less metallic high up to the right. The red line separates (high up to the right) the elements with definite non-metallic characteristics.
Thus Sodium (Na, atomic number 11) and Lithium (Li, atomic number 3) are metals, while Oxygen (O, atomic numer 8), Bromine (Br, atomic number 35) and Neon (Ne, atomic number 10) are non-metals.
At the same time Sodium (Na, 11) is more metallic than Lithium (Li, 3) and Manganese (Mg, 12), but less metallic than Caesium (Cs, 55).

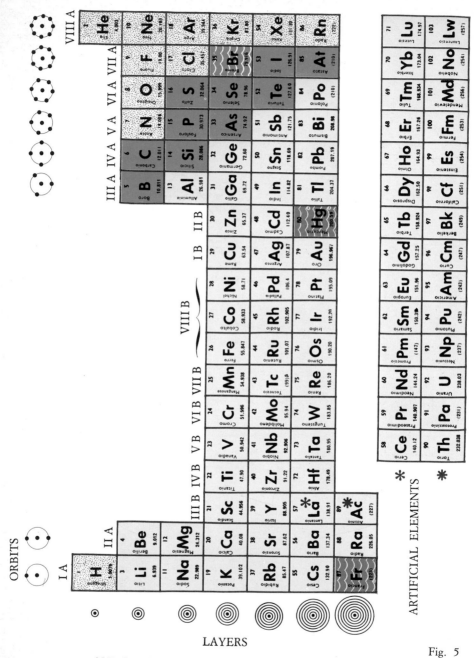

ORBITS

LAYERS

ARTIFICIAL ELEMENTS

Fig. 5

N.B. See also table on pages 245-6.

5) The majority of the elements are *solids* (for example, Silver). Only 11 are *gaseous* (for example, Oxygen) and only 2 *liquids,* mercury and Bromine.

6) Going from left to right the elements, although being always neutral, (the number of protons equals the number of electrons) have a larger number of protons in the nucleus (the atomic number increases). Because of this the force af attraction which the nucleus exerts on the electrons of the outermost layer progressively diminishes.

From this there follows a most important fact: in the periodic system the elements which radiate from the lower left hand corner tend to lose the electrons in the outermost layer more readily, transforming themselves into positive ions (cations): elements having this electro-positive tendency are the *metals*. On the other hand the elements which radiate from the top right hand corner tend to acquire new electrons in the outer layer and become negative ions (anions): elements having this electro-negative tendency are the *non-metals*. The notable exceptions are the noble gases which have no tendency either to capture or give up electrons, but are electrically stable. The reason for this stability is to found in the fact that they alone among all the existing elements, have 8 electrons in the outermost layer and are thus electrically saturated. (Helium, as we have already noted, having only a single layer is electrically saturated with 2 electrons).

FROM ELEMENTS TO COMPOUNDS: MOLECULES

As we have already noted, all the elements tend to saturate themselves electrically, that is, to form their outermost electronic layer on the model of the inert gases which have 8 electrons.

But how do they arrive at this?

They do it by giving up, capturing or sharing electrons with other elements, even at the cost of losing their freedom and tying themselves to other atoms. This is how "compounds" come into being.

Let us examine *how and in what number* the atoms of elements link up to form the molecules of compounds.

1) How atoms link up with each other (chemical linkage)

There are two principal kinds of chemical linkage: ionic linkage and co-valent linkage.

a) Ionic linkage

We already know that an atom can lose or acquire electrons in the outermost layer transforming itself into "ions" (positive or negative respectively). For example, Sodium (Na) in the presence of Chlorine (Cl) tends to let the latter capture its single electron in the outermost layer, so that both achieve 8 electrons. The result is that Sodium becomes a Sodium ion with a positive charge (written Na$^+$) and Chlorine becomes a Chlorine ion with a negative charge (Cl$^-$).

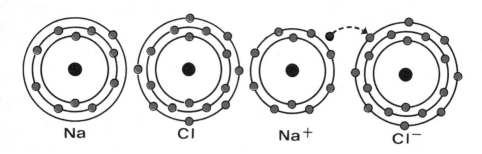

Fig. 6

But we know that electric charges of different signs are attracted to each other (see footnote 1 on page 15), and the ions are in fact atoms with electrical charges of different signs: from this arises the reciprocal attraction which links them together to give the molecule (electrically neutral) of Sodium Chloride (NaCl) which is commonly known as table salt.

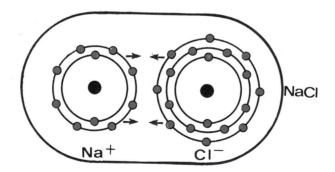

Fig. 7

The ionic linkage combines metals with non-metals. In fact, it is evident that this kind of linkage can arise only between positive ions and negative ions, that is, with ions derived from electro-positive elements (metals) and ions derived from electro-negative elements (non-metals).

The ionic linkage thus gives place to neutral molecules, but with an internal arrangement of positive and negative electric charges: such molecules are therefore said to be "polarized", that is, orientable like a magnetic needle.

b) The atomic or co-valency linkage

When two atoms can only give ions of the same sign (such as between atoms of the same element or between atoms of two non-metals), the link cannot be of the ionic kind when there are no ions of opposite electric charges.

These atoms, to achieve the octet (8 electrons), each provide a common electron (**known** as a co-valent bond) linking themselves together not by electrical attraction but by means of the atom itself: for this reason such a linkage is called "atomic" or, taking into account that each of the atoms contributes to the formation of the octet with the same number of electrons, "co-valent".

Thus two atoms of Fluorine (F) (7 electrons in the outermost layer), combine by sharing in common one electron each so as to form the "co-valent bond" that links them together in the molecule of Fluorine (F_2).

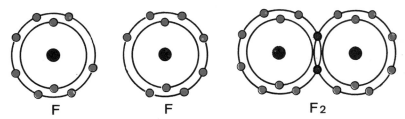

F   F   F₂

Fig. 8

Sometimes the atoms share more than one electron each, thus forming double or triple bonds. This is the case of the link between an atom of Carbon (C) (4 electrons in the outer orbit) and two atoms of Oxygen (O) (6 electrons in the outer orbit) to form the molecule of Carbon dioxide (CO_2). Each atom of Oxygen is attached to the Carbon atom with two co-valent bonds, while the atom of Carbon is attached to the atoms of Oxygen with four bonds (four co-valencies).

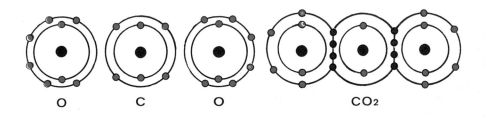

O   C   O   CO₂

Fig. 9

This link attaches together two atoms of the same element or two or more atoms of non-metals, and applies to the majority of chemical compounds.

Since these molecules do not have opposite electric charges internally, they are classed as non-polarized.

2) In what number atoms form a linkage (Valency)

As we have seen, all atoms tend to assume a stable structure forming an octet (8 electrons) in the outer shell. To do this they lose or capture (in ionic linkage) or share in common (in co-valent linkage) some of their electrons.

The *valency* of an atom is in fact the number of electrons that are involved to achieve the union with another atom.

Thus in the example on page 20, the atom of Sodium (Na) surrenders one electron to the atom of Chlorine (Cl); hence both Sodium and Chlorine have a valency of 1.

Two atoms of Fluorine (page 21) unite by sharing in common one electron each (1 bond): this means that Fluorine has a valency of 1.

The atom of Carbon (page 21) in uniting with Oxygen, shares two electrons with an atom of Oxygen (2 bonds) and another two another atom of Oxygen (another 2 bonds): this means that Carbon has a valency of 4 and Oxygen a valency of 2.

CLASSIFICATION OF COMPOUNDS

All chemical compounds are usually divided into two categories: organic and inorganic.

1) Organic compounds

These are the compounds which are connected with living organisms both vegetable and animal. In them the element of Carbon is always present.

There are an enormous number of organic compounds (almost three-quarters of known compounds) but they are of little interest to our present subject, and for this reason we shall say no more about them.

2) Inorganic compounds

These comprise all other compounds.
They may divided into three groups:

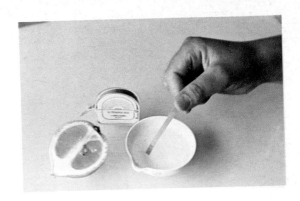

Fig. 10

a) *Acid compounds:* that is, which have a sour taste and turn blue litmus paper red.

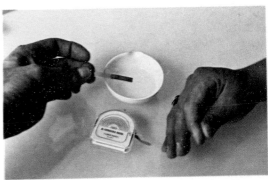

Fig. 11

b) *Alkaline compounds:* they are slippery to the touch and turn red litmus paper blue.

Fig. 12

c) *Neutral compounds;* that is, which are neither acid nor alkaline and make no change to the colour of litmus paper.

A - Acid compounds arise from the combination of a *NON-METAL* with *Oxygen* and take the name of *ANHYDRIDES.*

$$S \quad + \quad O_3 \quad = \quad SO_3$$
(Sulphur)   (Oxygen)   (Anhydride of Sulphuric acid)

If the *Anhydride* is combined with *water* is becomes an ACID

$$SO_3 \quad + \quad H_2O \quad = \quad H_2SO_4$$
(Water)   (Sulphuric acid)

NB — Five non-metals (Fluorine, Chlorine, Bromine, Iodine and Sulphur) form acids combining directly with Hydrogen.

$$Cl \quad + \quad H \quad = \quad HCl$$
(Chlorine)   (Hydrogen)   (Hydrochloric acid)

The basic characteristic of acids is that of containing in their molecule one or more atoms of Hydrogen that can readily be replaced by metals, freeing positive ions H^+ called Hydrogen ions. These H^+ ions are also liberated when acids are dissolved in water, increasing the concentration of H^+ ions in the water.

B - Base compounds result from the combination of a metal with *oxygen* and take the name of oxides.

$$K_2 \quad + \quad O \quad = \quad K_2O$$
(Potassium)   (Oxygen)   (Potassium oxide)

If *the oxide* is combined with *water,* a hydroxide (or base) is formed

$$K_2O \quad + \quad H_2O \quad = \quad 2 \, KOH$$
(Potassium oxide)   (Water)   (Potassium hydroxide)

The basic characteristic of Hydroxides (or bases) is that of containing in their molecule one or more groups OH^-, called hydroxyls, that are liberated when the hydroxides are dissolved in water, thereby increasing the concentration of OH^- ions in the water.

C - Compounds with neutral reaction result from the combination of an *acid* with a *hydroxide* and take the name of *SALTS.*
Such a reaction takes place by the substitution of the Hydrogen (H) of the acid with the metal of the hydroxide. The H^+ liberated from the acid combines with the OH^- liberated from the hydroxide forming water (H_2O) and also liberating heat.

24

$$H_2SO_4 \quad + \quad Ca(OH)_2 \quad = \quad CaSO_4 \quad + \; 2\,H_2O$$
$$\text{(Sulphuric acid)} \quad \text{(Calcium hydroxide)} \quad \text{(Calcium sulphate)} \quad \text{(Water)}$$

NB. - The salts are not always neutral: if the salt contains an H^+ from the acid, it has an *acid reaction;* if it contains an OH^- group from the hydroxide, it becomes *a base salt.*

Such salts are frequently used in photography and their greater or lesser degree of acidity or alkalinity is expressed as their pH value as explained below.

MEASUREMENT OF THE ACIDITY OR ALKALINITY OF A SOLUTION = pH

The pH is an index that varies from O to 14 which measures the excess of H^+ or of OH^- in solutions, that is, their acidity or alkalinity on the following basis:

$$\text{pH} \quad 0 \;=\; \text{maximum acidity}$$
$$\text{pH} \quad 7 \;=\; \text{neutrality}$$
$$\text{pH} \; 14 \;=\; \text{maximum alkalinity}$$

A simple method for measuring the pH is the use of "universal indicators". These are strips of paper impregnated with a special substance which on contact with the solution takes on different colours according to the acidity or alkalinity of the solution.

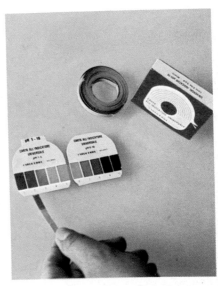

Fig. 13

A comparison with a coloured scale provided with the pH paper, will give the pH value.

"Universal indicators" are available for a range pH 1 to pH 11 and for more exact measurement, for example, from pH 6 to pH 8.

THE REACTIONS OF OXIDATION-REDUCTION

The reactions of oxidation-reduction are of particular importance in the field of photography.

When two elements react between themselves to form compounds or compounds react to separate into the elements of which they are composed, there is a passage of electrons from one reagent to the other. This passage of electrons is called oxidation-reduction. *Oxidation is the loss of electrons, and reduction is the capture of electrons.*

The reagent that *loses* electrons is the *object of reduction* which after the reaction becomes *oxidised.*

The reagent that *gains* electrons is the *object of oxidation* which, after the reaction, becomes *reduced.*

Fig. 14

It is clear that there cannot be oxidation without accompangng reduction.

In reactions of oxidation-reduction the displacement of electrons takes place from the atom less "hungry" for electrons to that which is more "hungry", by which it is possible to predict on the basis of elements of decreasing hunger which will be the object of oxidation and which the object of reduction.

A series known as the *"electro-chemical"* series gives the elements in increasing order of electron avidity.

Li; K; Ba; Sr; Ca; Na; Mg; Al; Mn; Zn; Cr; Fe; Co; Ni; Sn; Pb; H; Cu; I; Ag; Br; Pt; Cl; Au; F.

For example, in the reaction between Sodium (Na) and Chlorine (Cl), it is the Chlorine which robs electrons from Sodium, oxidising it and becoming reduced. In the same way, Bromine (Br) will oxidise Silver (Ag) becoming itself reduced.

THE PHOTOGRAPHIC EMULSION

1) THE STRUCTURE OF PHOTOGRAPHIC FILMS

There are certain salts of silver (the silver halides) which have a special characteristic: when they are struck by light their chemical composition undergoes a modification (photochemical effect): the small crystals of the silver halides which are transparent are transformed into opaque metallic silver.

The silver halide crystals are dispersed in gelatin forming a mixture which is known as a "photographic emulsion".

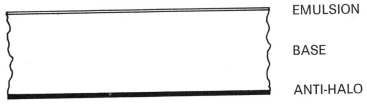

EMULSION

BASE

ANTI-HALO

Fig. 15

This emulsion is coated in a thin layer on sheets of glass or transparent film which provide a "base" for it.

The base of the film is usually composed of cellulose acetate or polyester. The latter has a very high dimensional stability, close to that of glass, and is indispensible when the images of two or more films are required to be superimposed in perfect registration.

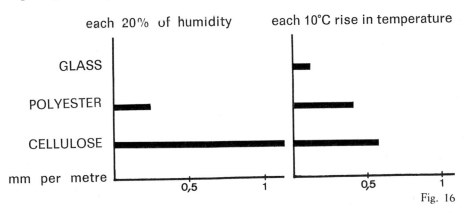

Fig. 16

The two graphs show the dimensional stability of different bases with variations of humidity and temperature.

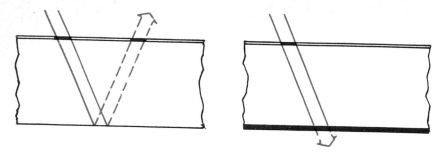

Fig. 17

The base is generally coated on the underside with a layer of coloured substance (called "anti-halo") which absorbs light rays so as to prevent them reflecting back into the emulsion to form a double image or "halo".

2) THE PREPARATION OF A PHOTOGRAPHIC EMULSION

The silver halides are the salts which silver forms when it combines chemically with the four metalloids of the group of halogens (= generators of salts): fluorine, chlorine, bromine and iodine (see the Periodic Table of elements on page 18).

Of these salts, Silver Bromide is the most sensitive to light and is thus the most commonly used in photographic emulsions.

Silver Bromide is usually made by mixing Silver Nitrate with an alkaline halide — for example, Potassium Bromide — in a aqueous solution of gelatin.

$$AgNO_3 \quad + \quad KBr \quad = \quad AgBr \quad + \quad KNO_3$$
(Silver nitrate) (Potassium Bromide) (Silver Bromide) (Potassium nitrate - soluble)

In the reaction the Silver Bromide precipitates in the form of crystals which can be left to "grow" more or less in size (= physical ripening), after which the Potassium Nitrate, being soluble, is removed by washing with water.

At a later stage, with the object of increasing the light-sensitivity of the Silver Bromide, certain chemical substances are added (in particular Sulphur compound) which attach at various points to the sides of the crystals of Silver Bromide with the formation of "impurities" of Silver Sulphide (= chemical ripening).

These deposits of Silver Sulphide are, as we shall see, a principal ingredient in the increase of the sensitivity of the emulsion and are therefore called "sensitivity specks".

Diagram showing the preparation of the photographic emulsion.

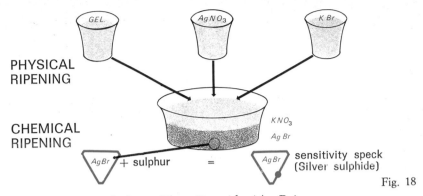

PHYSICAL
RIPENING

CHEMICAL
RIPENING

Fig. 18

Now another word about Silver Bromide (Ag Br).

a) Silver Bromide is a salt whose molecule is formed by means of an ionic link (see page 20). In fact Silver is a metal with a single electron in its outer orbit and Bromine a non-metal with 7 electrons in its outer orbit.

In the presence of silver, bromine will tend to capture the single electron to form the octet (8 electrons), thus creating two ions:

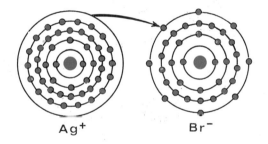

Fig. 19

as these are of different sign, they unit (with ionic linkage) to form the salt Silver Bromide (Ag Br), which is both electrically and chemically neutral.

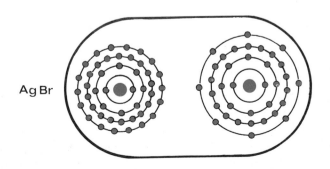

Fig. 20

29

Fig. 21

b) Silver Bromide normally crystallises to form a cube, but when crystallisation takes place in gelatin, the crystals assume an octahedral form, though for reasons which are as yet unknown.

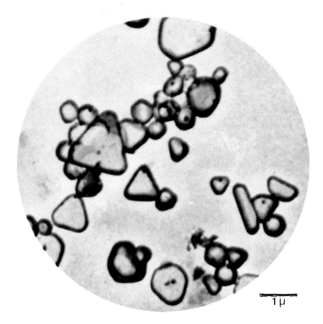

Fig. 22

The figure shows a photograph of Silver Bromide crystals taken with an electron microscope. The horizontal line indicates the length of 1 micron ($\mu = \dfrac{1}{1000}$ of a millimetre).

3) THE EXPOSURE AND FORMATION OF THE LATENT IMAGE

Light, whose intimate nature still remains a mystery, consists of particles of energy (called quanta of light or photons) which are propagated by a wave motion.

When a photon of energy (symbol hv) strikes a crystal of Silver Bromide, an electron on the periphery (symbol e^-) of the ion Br^- may be pushed out of its orbit, transforming the ion Br^- to the atom Br, which then migrates out of the crystal.

$$Br^- \quad + \quad hv \quad = \quad Br \quad + \quad e^-$$
(Bromine ion) (photon) (Bromine atom) (electron)

Nothing equivalent can occur in the ion Ag^+ since its positive charge strongly holds back the outer electrons. It may happen that the ion Ag^+ (positive) is near an electron e^- (negative) and by attracting it, reduces itself to an atom of metallic silver (Ag).

$$Ag^+ \quad + \quad e^- \quad = \quad Ag$$
(Silver ion) (electron) (atom of metallic silver)

The atom of metallic silver then wanders in the crystal until it meets an impurity (for example, silver sulphide) or a defect in the crystal where it remains fixed.

The impurity or crystal defects are what we have already referred to as the sensitivity specks and the atoms of metallic silver which accumulate there (ideally in groups of 3 to 12 specks) constitute the "latent image".

FORMATION OF THE LATENT IMAGE

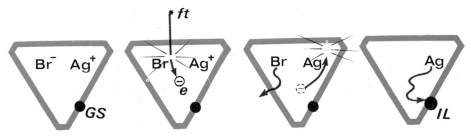

Fig. 23

Silver bromide crystal with «sensitivity speck» (G.S.).	The photon (ft) oxidises the ion Br^- to the atom Br, liberating an electron e^-.	The electron reduces the ion Ag^+ to the atom Ag of metallic silver. The atom of Br wanders in the crystal until it escapes.	The atom of metallic silver wanders in the crystal until it locates a sensitivity speck where it fixes itself giving rise to the «latent image» (I.L.).

4) DEVELOPMENT

The purpose of development is to reduce to metallic silver *all* the ions Ag^+ of the crystals in which the latent image has been formed, so as to obtain a complete blackening of those crystals struck by light and at the same time leaving those not exposed intact.

Since the reduction is operated by the electrons, it is necessary to bring the crystals in contact with a strongly "base" substance (see page 24 and 26) which is capable of giving up its own eletrons: this substance is called the developer.

Developers are, therefore, no more than "reducers" which by continuing the reduction started by the electrons during the exposure to light, "develop" or intensify the latent image.

It is clear that developers release electrons to the Ag^+ ions, and in the process of reducing these, "oxidise" themselves: we are thus dealing with an "oxidation-reduction" reaction (see page 26).

$$\text{Reducer} \overset{e^-}{\;} + \;\; Ag^+ \;\; \rightarrow \;\; \underset{Ag}{\underline{Ag^+ + e^-}} \;\; + \;\; \text{Oxidised reducer}$$

(Developer)　　(Silver ion)　　(Metallic silver)

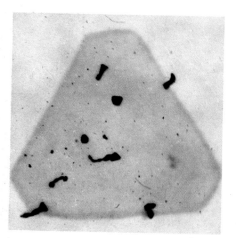 

Fig. 24

The illustration on the left shows a crystal of Silver Bromide (AgBr) taken with an electron microscope after exposure to light and with a brief development (the location of "sensitivity specks" can be seen). On the right, the same crystal is shown after full development. It can be seen how the metallic silver has grown in a filamentary form from the "speck".

The dispersion of the silver bromide crystals in gelatin prevents the developer from reducing those crystals which have been unaffected by the exposure to light.

However, with prolonged development even the unaffected crystals will be reduced thus forming a uniform "fog" over the whole emulsion (chemical fog).

THE ACTION OF DEVELOPMENT

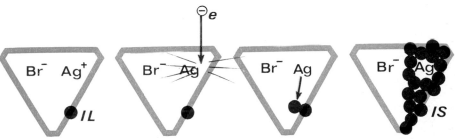

Fig. 25

Crystal with « latent image » (I. L.) formed from an atom of metallic silver.

The developer furnishes an electron (e) which reduces the ion Ag+ to Ag (metallic silver).

The atom of metallic silver joins the others which already form the latent image.

After development, only half of the ions of the crystal (that is, all the Ag+ and not the Br—) are reduced to visible metallic silver or the developed image (I.S.).

NB. - The reaction so far described constitutes *"Chemical Development"*. At the same times as the chemical development, a secondary phenomenon of silver reduction occurs to a greater or lesser extent which is inappropriately called *"Physical Development"*.

This physical development occurs as follows: in almost all developers there is a certain amount of *Sodium Suphite*. This salt has the property of dissolving the crystals of silver bromide to form a complex of silver and sulphite. When dissolved in the developer, the latter reduces it to form metallic silver.

This metallic silver originates from *all* the silver bromide crystals, including those not affected by exposure, (unlike that which arises as we already know from "chemical" development which arises only from crystals affected by light) but fortunately, this silver also has a preference to deposit itself on the exposed crystals thereby helping the growth of the photographic image.

Excessive physical development could, however, deposit metalic silver at random causing fogging of the emulsion (physical fog).

In photomechanical processes two main types of developer are used:

a) *High contrast developers* so as to obtain only blacks and very clear whites.

They are used with "lith" films of steep gradation for the reproduction of line and half-tones (dot images).

The most common is the *paraformaldehyde* developer and as an example we give the formula for the Ferrania-3M (R-35).

Solution A

Anhydrous sodium sulphite	30	gm. (cryst. 60 gm.)
Boric acid	7.5	gm.
Hydroquinone	22.5	gm.
Potassium bromide	1.5	gm.
Water to make	750	cc.

Solution B

Potassium metabisulphite	2.6	gm.
Sodium sulphite (anhydrous)	1	gm. (cryst. 2 gm.)
Paraformaldehyde	7.5	gm.
Water to make	750	cc.

It is used by mixing 3 parts of A and 1 part of B immediately before use: develop for 2½ to 3 minutes at 20°C.

The pH value is about 6 for solution A and 10 for solution B.

b) *Low contrast developers* so as to obtain a full range greys. They are used with soft-gradation films for the reproduction of continuous tone originals.

The most common of these are *the metol-hydroquinone developers* of which the Ferrania-3M (R-45) is typical.

Metol	1.5	gm.
Sodium sulphite (anhyd.)	50	gm. (cryst. 100 gm.)
Hydroquinone	9	gm.
Sodium carbonate (anhyd.)	25	gm. (cryst. 67 gm.)
Potassium bromide	4	gm.
Water to make	1000	cc.

Develop for 3 to 5 minutes at 20°C.

The pH value is around 9 to 10.

5) THE STOP BATH

The purpose of the stop bath is to stop any further reducing action of the developer.

This bath is usually no more than a dilute solution of glacial acetic acid (on the basis of 30 cc. of acid to 1000 cc. of water) which immediately neutralises the alkalinity of the developer remaining in the film and thus prevents any further action.

The pH value of the stop bath should be around 3 to 4.

6) FIXATION

This has the purpose of dissolving all the undeveloped silver bromide crystals in the emulsion so as to leave only the silver image.

The fixing bath is usually a solution of sodium thiosulphate (also called sodium hyposulphite or "hypo") which has the formula $Na_2S_2O_3$.

The process involves the formation of complex silver sodium salts which are soluble in water and may thus be removed by giving the film a thorough washing in running water so as to eliminate all the by-products of fixation.

Photomechanical processes make use of two main types of fixer:

a) *Normal fixer*

Fixing is complete in about 10 minutes.

Sodium thiosulphate (anhyd.)	255 gm. (cryst. 400 gm.)
Sodium bisulphite (anhyd.)	50 gm.
Water to make	1000 cc.

The chemicals are dissolved in the order given in about two-thirds of the water at about 30°C. When dissolved the volume is made up to 1 litre with cold water.

The pH value should be maintained from 3.5 to 6.

b) *Rapid Fixer*

Fixing is complete in about 5 minutes.

Sodium thiosulphate (anhyd.)   160 gm. (cryst. 250 gm.)
Sodium bisulphite (anhyd.)   18 gm.
Ammonium chloride   40 gm.
Water to make   1000 cc.

The pH should be maintained within 3.5 to 5.

Summarising, the three stages (exposure, development and fixation) that give rise to the photographic image, can be shown diagrammatically:

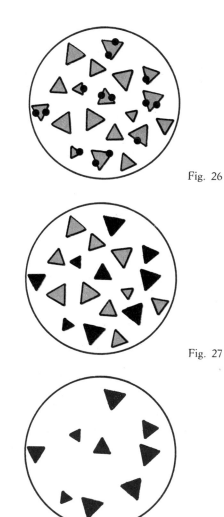

LATENT IMAGE

Fig. 26

DEVELOPED IMAGE

Fig. 27

FIXED IMAGE

Fig. 28

7) REDUCTION OF THE IMAGE

The object of reduction is partly to dissolve the metallic silver of the developed and fixed image so as to make it more transparent.

Photographic reducers are essentially "oxidising" substances which, subtracting electrons from metallic silver, transform it into positive silver ions Ag^+.

$$\text{Reducer} \quad + \quad Ag \quad \rightarrow \quad Ag - e^- \quad + \quad \text{reduced "reducer"}$$

From this oxidation-reduction reaction there follows the linkage of Ag^+ ions with the negative ions already contained in the reducer with the formation of a soluble silver salt.

In this way the metallic silver is partly dissolved so as to make the silver image less dense.

Photographic reducers are classified according to the effect they have on the contrast of the image:

a) *Subtractive reducers,* which lower all densities by an equal amount so that the contrast of the image is substantially unaltered.

area dissolved

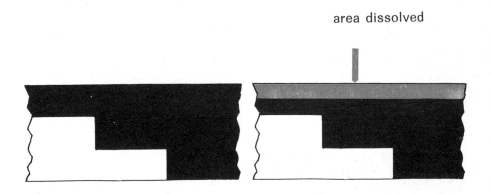

Fig. 29

A typical subtractive reducer is that known as Farmer's reducer which is prepared in the following way.

Solution A Sodium thiosulphate (anhyd.)   32 gm. (cryst. 50 gm.)
           Water to make   500 cc.
Solution B Potassium ferricyanide   5 gm.
           Water to make   500 cc.

The solutions are mixed immediately before use to make a suitable volume. A less vigorous action can be obtained by adding more water.

Fig. 30

Farmer's reducer can be applied locally by applying it in three stages:

1st - Using a small paint brush to apply Solution B to areas which are to be made less dense.
2nd - A wad of cotton wool soaked in solution A is applied over the area a few moments later.
3rd - The negative is thoroughly washed with running water.

b) *Proportional reducers,* which lower the contrast.

These react more strongly where there is more silver.

area dissolved

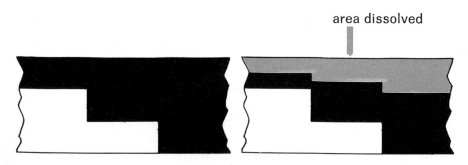

Fig. 31

The most important reducer of this kind is that known as a "permanganate reducer" which is made up as follows:

Potassium permanganate	0.5	gm.
Sulphuric acid (10% sol.)	10	cc.
Water to make	1000	cc.

As this solution is not stable, it must be prepared immediately before use. It leaves the gelatin stained a brownish colour and it is thus necessary to rinse the negative in a solution of sodium bisulphite:

Sodium bisulphite (anhyd.)	100	gm.
Water to make	1000	cc.

8) INTENSIFICATION

A photographic intensifier has the purpose of increasing the density of the image by depositing metal (Silver or Mercury) over the existing silver image. (Make sure the film does not suffer from fog, as this will also be increased: if necessary, the fog should first be reduced to a negligible level with Farmer's reducer).

The degree of intensification is usually proportional to the amount of silver already present, so that the darker areas are more strongly intensified than the clearer areas. This tends to increase the contrast.

Intensification of this kind is thus extremely useful for treating line and screen images which are too weak.

zone intensified

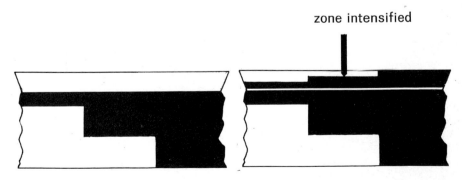

Fig. 32

The most common intensifier is that of *Mercuric Chloride* which is made up as follows:

Mercuric chloride	20 gm.
Sodium chloride	50 gm.
Hydrochloric acid (conc.)	5 cc.
Water to make	1000 cc.

The negative is immersed in the solution until the image takes on a whitish appearance and is then washed in running water for about 15 minutes. After this it is "developed" in a normal metol-hydroquinone developer.

N.B. - Effect on half-tone screen images

To reduce the area of the dots of a screen image (see page 100), it is preferable to use Farmer's reducer (page 38).

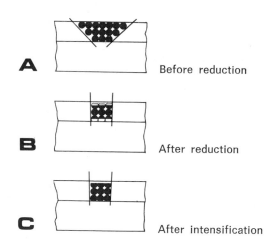

A   Before reduction

B   After reduction

C   After intensification

Fig. 33

Seen as an enlarged section, the half-tone dot (if the exposure and development have been exact) take the form of an agglomeration of silver grains shaped like an inverted trapezium (Fig. 33 A).

After reducing, the silver grains at the sides and above are oxidised and thus dissolved. Above and below the dot, some grains are only partly oxidised (Fig. 33 B). Whenever the reduction is pushed too far, not only will the area of the dot be diminished, but also its density. In this case it is necessary to treat the film with an intensifier so as to build up the partly dissolved grains (Fig. 33 C). A strengthening of the dots can also be obtained by making an inter-negative on a high contrast film (by contact), from which a new positive can be made.

CHAPTER 2

PHOTOMETRY

To be able to measure the effect of light upon a photographic emulsion, it is first necessary to measure the light itself. The brief study of photometry (= measurement of light) which follows, concerns the *intensity* of a light source, the *flux* of light that it emits, and the *illumination* received by the surrounding objects.

We can start our investigation by considering an ordinary lighted candle.

Fig. 34

The flame of a candle possesses a certain quantity of light (luminous intensity) which is propagated in all directions (luminous flux) and illuminates the surrounding objects (illumination).

These three factors form the basis of this chapter:

1) The light source (intensity and flux);
2) The objects illuminated (illumination);
3) The units by which these are measured.

In order to make the present chapter more understandable, we shall first consider some of the ideas about the geometry of the sphere. However, as these are not strictly indispensible, the reader may, if so inclined, skip the next few pages and continue on page 45.

GEOMETRY OF A SPHERE

1 - Surface and solid angle of a sphere

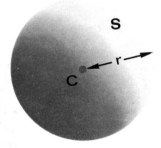

Fig. 35

We know from geometry that a sphere with a centre C and a radius r has a surface S

$$1) \quad S = 4\pi r^2$$

We can see that:
if $r = 1$ the surface is 4π
if $r = 2$ the surface is $4\pi \times 4$
if $r = 3$ the surface is $4\pi \times 9$
that is, the surface S varies with r^2, while 4π remains constant for spheres of whatever dimensions you may think of.

In fact, 4π is the measure of the "solid angle" which, radiating from the centre C in all directions gives rise to the sphere, and is — as can be seen from (1) — the constant ratio between the surface S of the sphere and the square of the radius.

$$2) \quad 4\pi = \frac{S}{r^2}$$

2 - Suface and solid angle of a sector of a sphere

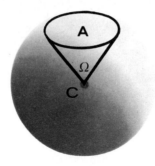

Fig. 36

Instead of considering the whole suface S of the sphere, let us take a part of the surface having an area A.

If from the circumference that delimits the area A we lower radii towards the centre C of the sphere, we shall obtain a cone (more precisely a spherical sector) with base A and vertex C whose solid angle (let us call it Ω) will be the constant ratio between A and r^2.

From (2) we deduce in fact that

$$3) \quad \Omega = \frac{A}{r^2}$$

and from this that:

$$4) \quad A = \Omega \cdot r^2$$

3 - Unit sphere, unit solid angle, Steradian

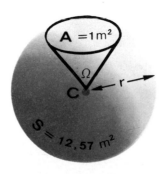

Fig. 37

We can now give unit values to the sphere:

if the radius r is 1 metre,

$$r = 1 \text{ mt}$$

the surface S of the sphere will be $4\pi\, r^2$, that is 12.57 x 1:

$$S = 12.57 \text{ mt}^2$$

and the unit area A will be the 12.57th part of S, that is

$$A = 1 \text{ mt}^2$$

Consequently the unit solid angle Ω will be 1. In fact from 3) we deduce

$$\underline{\Omega} = \frac{A}{r^2} = \frac{1}{1} = 1$$

The cone with base A and vertex C, derived from the unit solid angle Ω (corresponding to the 12.57th part of the solid angle of the sphere), is called the Steradian.

From this it follows that:

a) Every sphere, whatever its radius, is composed of 12.57 Steradians.

b) In the Steradian the solid angle Ω (always = 1) = $\dfrac{A}{r^2}$

c) In the Steradian, the spherical cap $A = \Omega\, r^2$
   and since $\Omega = 1$ $A = r^2$

PHOTOMETRY

1) THE LIGHT SOURCE

Let us start with an example.
A bar of iron heated in a flame becomes red hot and emits light.

Fig. 38

The light emitted is more intense where the temperature is highest and loses intensity gradually as the heat of the bar decreases.

Let us now assume that we are able to separate, at various points along the bar, small particles of incandescent iron, similar to small light sources (see footnote 1) of different luminous intensity.

Fig. 39

Each one of them will radiate a luminous *flux* in all directions in the form of a sphere, which will be so much the greater as the source is intense.

Footnote (1) Let us imagine that these particles are so small as to resemble points with practically no surface (that is, point sources).

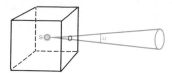

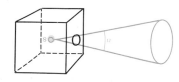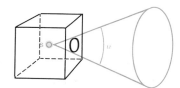

Fig. 40

Let us now place one of these luminous point sources S in a series of boxes, in one side of which there is an aperture progressively bigger in diameter (Fig. 40). It is obvious that the cone of light which comes from the aperture will have an angle Ω (see footnote 1) progressively bigger and that, in consequence, the quantity of light that comes from the box (that is the luminous flux) will be progressively greater.

In other words we can say that the *luminous flux* depends on two factors;

a) on the intensity of the source (Fig. 39): the greater the intensity of the source, the greater the flux.

b) on the angle of radiation (Fig. 40): the greater the angle of radiation, the greater the flux.

We can now write that the flux Φ is equal to the product of the intensity I of the source and the angle Ω of radiation (see footnote 2):

5) $\Phi = I \times \Omega$

From this formula we deduce that:

6) $I = \dfrac{\Phi}{\Omega}$

In fact if the angle of radiation is constant (same aperture), it is necessary to vary the intensity of the source (Fig. 41) in order to vary the flux: increasing the intensity of the source increases the flux; reducing the intensity reduces the flux. (I and Φ are directly proportional).

Footnote (1) The reference here is the idea of a « solid angle » as defined in the pages referring to the geometry of the sphere (page 43).

Footnote (2) We are omitting for the present the duration of the luminous flux, that is the *time* factor, which we shall include when come to deal with the exposure (page 50).

46

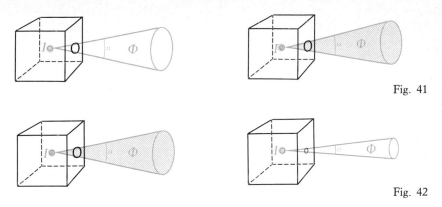

Fig. 41

Fig. 42

On the other hand, to maintain a constant flux of light Φ, any increase in the luminous intensity I must be offset by a corresponding reduction in the angle of radiation Ω (smaller aperture - fig. 42), and vice versa. (Thus I and Ω are inversely proportional).

NB. - The Brightness (or brilliance) of a source

If, instead of being a point, the source were of *extended area* while still retaining the same intensity I (or quantity of light), it would lose in brilliance B.

7) $$ B = \frac{I}{S} $$

2) OBJECTS ILLUMINATED

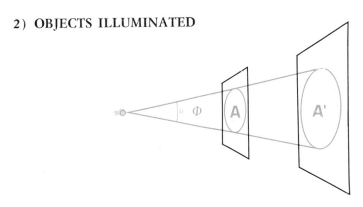

Fig. 43

The luminous flux Φ which radiates from a luminous source illuminates surrounding objects: this phenomenon is called *illumination*.

47

The luminous flux Φ that is emitted from the source S (see Fig. 43) is allowed to fall on the surface A of a sheet of paper placed at a certain distance from the source (1st position). Every single point of A will receive a certain illumination. If we now move the paper farther from the source (2nd position), the flux Φ will illuminate a greater area A'. The flux Φ is still the same, but in A' is spread over a greater surface: from this we can deduce that every single point on surface A' will be less illuminated than every single point on surface A.

From these observations we can conclude that the illumination L on every point of the surface A (or A') is given by the total luminous flux Φ divided by the surface illuminated:

$$8) \quad L = \frac{\Phi}{A}$$

If we now wish to know what illumination is produced by a light source of given intensity placed at a given distance from an object (this is what interests us most), it is sufficient to substitute in formula 8 the flux Φ with I x Ω (on the basis of formula 5), and the surface A with Ω x r^2 (on the basis of formula 4) to arrive at

$$L = \frac{I \cdot \Omega}{\Omega \cdot r^2}$$

and by cancelling out Ω

$$9) \quad L = \frac{I}{r^2}$$

which is the fundamental ratio in photometry. It tells us that the illumination on every point of a surface *is directly proportional to the intensity of the source and inversely proportional to the square of the distance from the source.*

This law can be represented diagrammatically - Fig. 44.

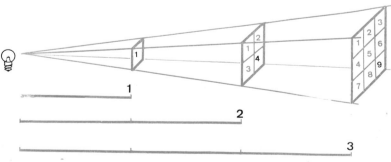

Fig. 44

3) UNITS OF LIGHT MEASUREMENT

Before listing the units of light measurement that are in common use at the present time, it is useful to recall to mind the idea of a *unit solid angle* (Steradian) given on page 43 since from now on we shall be referring to the Steradian (Ω) whose values (Ω; A; r) are all equal to 1 ($\Omega = 1$ St; $A = 1$ metre^2; $r = 1$ metre).

We shall begin with the unit of *luminous intensity* (I) of the source (to- which we shall add that of its *brightness* (B)), then we shall give that of the *luminous flux,* (Φ) and finally that of *illumination* (L).

1. Unit of luminous intensity (I) of the source: the Candela (Cd)

The candela is the fundamental unit of measurement of light and has been chosen by scientists on an international basis. It does not differ very much from the standard wax candle which was previously adopted for this purpose.

2. Unit of brightness (B) of the source: the Lambert (Lamb.)

The brightness or luminance of an extensive or diffuse source (that is, not a point source) is expressed in terms of candelas per unit area. It may be stated as candelas per square metre but for a light source this is an excessively large unit. The lambert (L) is the brightness of a source or reflecting surface emitting or reflecting 1/Pi candelas per square centimetre.

3. Unit of luminous flux (Φ): the Lumen (lm.)

The lumen is the amount of light emitted in unit solid angle (St.) from a point source having an intensity of 1 candela.

$$lm = cd \times \Omega \qquad \text{(see formula 5)}$$

4. Unit of illumination (L) received by an object: the Lux

A lux is the illumination received by a surface at a distance of 1 metre from a source having a luminous intensity of 1 candela. It is equal to a flux of 1 lumen evenly distributed over an area of 1 square metre.

$$lux = \frac{I}{m^2} \qquad \text{(see formula 8)}$$

or, in other words, it is the illumination given by 1 candela on an object at a distance of 1 metre. For this reason lux is often referred to as a metre-candela.

EXPOSURE AND THE RECIPROCITY LAW

Exposure

Since a photographic emulsion can be thought of as "storing" the effect of light energy, we must take account not only of the intensity of the light striking the film, but also of the time that it is allowed to reach the film. The fundamental unit of photographic exposure is the lux-second (metre-candela second). It it the light energy of 1 lux acting for a time of 1 second or its equivalent, eg. 1/10 lux acting for 10 seconds, 5 lux for 1/5 second and so forth.

$$E = \text{lux x sec.}$$

which is the unit of measurement of the exposure.

Naturally to arrive at the effective exposure which an emulsion receives in a given time, it is necessary to multiply the illumination (in lux) by the time (in seconds).

$$E = I \times t$$

For example, if an illumination of 3 lux is applied for 10 seconds, the total value of the Exposure (E) will be given by the product of the lux by the time in seconds

$$E = 3 \text{ lux x 10 seconds} = 30 \text{ lux}$$

The Reciprocity Law (of Bunsen and Roscoe)

According to this law the blackening of a photographic emulsion depends on the total luminous energy (E) employed, independently of the values of I and t.

On this basis the same amount of blackening will take place exposing a photographic emulsion to an exposure

$$
\begin{aligned}
E &= &1 \text{ lux x } 1/10 \text{ sec.} \\
\text{or } E &= &10 \text{ lux x } 1/100 \text{ sec.} \\
\text{or } E &= &1/10 \text{ lux x 1 sec.}
\end{aligned}
$$

since the product of I and t is constant.

Failure of the reciprocity law (Schwarzschild effect)

However, when the value of I becomes too small or that of t too long, the "reciprocity law" fails, and to have the same blackening of the film, it becomes necessary to apply a special increase in the total exposure.

In this case the general equation

$$E = I \times t$$

must be modified to

$$E = I \times t^k$$

where the exponent k represents the power to which t is increased to obtain a correct exposure.

The exponent k is a characteristic of each type of emulsion and, when necessary, is indicated by the manufacturer of the film.

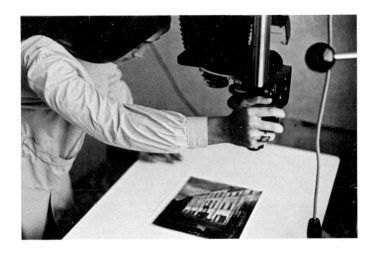

CHAPTER 3

DENSITOMETRY

Densitometry is the technique of measuring the photochemical effect (blackening) caused by light on a photographic emulsion.

As usual, we begin by recalling aspects of mathematics which concern ratios and proportions, since these have a direct bearing on our subject.

A - **RATIOS AND PROPORTIONS**

The ratio is a quotient, that is, the result of dividing one number by another.

To indicate the ratio between A and B, we write:

$$A : B \text{ or } \frac{A}{B}$$

and read "A is to B" or "A divided by B".

The second method by which we can represent a ratio ($\frac{A}{B}$) is as a "fraction". In this the number (A) over the line of the fraction is called the "numerator" and that under (B) "denominator". By carrying out the division we obtain the result which can be a whole number or a decimal.

Examples:

$$\frac{6}{3} = 2 \text{ (whole number)}$$

$$\frac{18}{9} = 2 \text{ (whole number)}$$

$$\frac{4}{5} = 0.8 \text{ (decimal number)}$$

It can be seen that two different fractions, $\frac{6}{3}$ and $\frac{18}{9}$ give the same result of 2 *because* the values of the second fraction 18 and 9 are none other than the values of the first 6 and 3 multiplied by the same number:

$$\frac{6 \times 3}{3 \times 3} = \frac{18}{9} = 2$$

We can thus say that, in multiplying (or dividing) both the numerator and denominator by the same number, the result of the division (that is the value of the ratio) does not alter.

The proportion is the equality between two ratios or fractions. If we write

$$\frac{6}{3} = \frac{18}{9}$$

we have written a "proportion", that is an "equation" between two ratios.

We could also write this proportion like this:

$$6 : 3 = 18 : 9$$

and read it by saying "6 is to 3 as 18 is to 9".

In this proportion 6 and 9 (which are the outer terms), can be called the "extremes", while 3 and 18 (which are the inner terms), can be called the "middles".

We are now in a position to state one of the fundamental properties of proportions: "The product of the middles is equal to the product of the extremes"

that is, if $6 : 3 = 18 : 9$

then $6 \times 9 = 3 \times 18$

in fact $54 = 54$

If the ratios are written as fractions

$$\frac{6}{3} = \frac{18}{9}$$

6 and 9 are the extremes, 3 and 18 the middles, and their equality can be written on the basis of "cross multiplication"

$$\frac{6}{3} \times \frac{18}{9}$$

$$6 \times 9 = 3 \times 18$$

This method is also called "crossed exchange" since by a crosswise exchange of the terms, the value of the proportion does not change.

In fact the proportion

$$\frac{6}{3} = \frac{18}{9}$$

remains the same if it is written with the "extremes" changed over

$$\frac{9}{3} = \frac{18}{6}$$

or changing over the "middles"

$$\frac{6}{18} = \frac{3}{9}$$

or changing over both

$$\frac{9}{18} = \frac{3}{6}$$

Now remember that a number does not change it value when it is divided by 1. In fact

$$10 : 1 = 10 \qquad 0,8 : 1 = 0,8$$

which written as fractions become

$$\frac{10}{1} = 10 \qquad \frac{0,8}{1} = 0,8$$

If we now wish to find out the value of an "unknown" in the equation

$$6 = \frac{12}{a}$$

we can imagine seeing it written

$$\frac{6}{1} = \frac{12}{a}$$

which, by changing over the extremes, becomes

$$\frac{a}{1} = \frac{12}{6} = 2 \quad \text{that is, } a = \frac{12}{6} = 2$$

Or wishing to know the value of the unknown a in the equation

$$4 = \frac{a}{6}$$

we can imagine it written

$$\frac{4}{1} = \frac{a}{6}$$

which, cross multiplied, will give

$$1 \times a = 4 \times 6$$
that is, $\quad a = 4 \times 6 = 24$

From these examples we obtain the formula for isolating an unknown value in an equation:

$$A = \frac{B}{C} \qquad C = \frac{B}{A} \qquad B = A \times C$$

B - LOGARITHMS

"Common logarithms" have a base of 10 and the logarithm of a number is the exponent to give to 10 to have that number.

Therefore, if $\qquad 10^2 = 100$

2 is the common logarithm of 100

and is written $\qquad 2 = \log_{10} 100$

or more simply $\qquad 2 = \log 100$

The original number (in this case 100) is called the "antilogarithm",

thus $\qquad 100 = $ antilog. 2

The logarithms of numbers less than 1 (fractional numbers) are given a negative sign

$$\log. \frac{1}{10} = -1; \quad \log. \frac{1}{100} = -2; \quad \log. \frac{1}{1000} = -3$$

and are usually written with the minus sign above the log which is then generally referred to as "bar".

Thus, $\overline{1}$ (bar one): $\overline{2}$ (bar two) and so on.

The following simplified table gives the logarithms for numbers from 1 to 1000 which is more than sufficient for sensitometric purposes.

54

TABLE OF COMMON LOGARITHMS FOR NUMBERS FROM 1 TO 1000

Number (Antilog)	Log	Number (Antilog)	Log	Number (Antilog)	Log
1.00	0.00	11.2	1.05	112	2.05
1.12	0.05	12.6	1.10	126	2.10
1.26	0.10	14.1	1.15	141	2.15
1.41	0.15	15.8	1.20	158	2.20
1.58	0.20	17.8	1.25	178	2.25
1.78	0.25	20.0	1.30	200	2.30
2.00	0.30	22.4	1.35	224	2.35
2.24	0.35	25.1	1.40	251	2.40
2.51	0.40	28.2	1.45	282	2.45
2.82	0.45	31.6	1.50	316	2.50
3.16	0.50	35.5	1.55	355	2.55
3.55	0.55	39.8	1.60	398	2.60
3.98	0.60	44.7	1.65	447	2.65
4.47	0.65	50.1	1.70	501	2.70
5.01	0.70	56.2	1.75	562	2.75
5.62	0.75	63.1	1.80	631	2.80
6.31	0.80	70.8	1.85	708	2.85
7.08	0.85	79.4	1.90	794	2.90
7.94	0.90	89.1	1.95	891	2.95
8.91	0.95	100.0	2.00	1000	3.00
10.00	1.00				

The use of logarithms is extremely useful because it simplifies calculations by reducing multiplications to simple additions and divisions to simple subtractions, as shown below:

1) The logarithm of a product is equal to the sum of the logarithms of every factor.

$$\log (a \times b) = \log a + \log b$$

If we wish to multiply 10 x 10 we proceed as follows:

$\log (10 \times 10) = \log 10 + \log 10 = 1 + 1 = 2$
antilog of 2 = 100
that is 10 × 10 = 100

2) The logarithm of a quotient is equal to the difference between the logarithm of number to be divided and logarithm of the divisor:

   log (a ÷ b) = log a — log b

For example, if we wish to divide 316 by 2, we proceed as follows:

   log (316 ÷ 2) = log 316 — log 2 = 2.50 — 0.30 = 2.20
   antilog of 2.20 = 158
   that is, 316 ÷ 2 = 158

C - GRAPHS OF MATHEMATICAL FUNCTIONS

It is useful to make a brief reference to the use of graphs to represent simple mathematical functions. This will help us to understand the nature of the characteristic curve of an emulsion which we shall have occasion to speak of later.

1) Coordinates

Let us consider two straight lines x and y which meet each other at O, their point of origin.

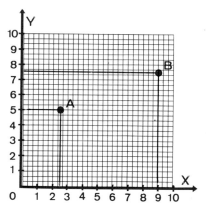

Fig. 45

The line x is called the horizontal axis and the line y the vertical axis. The two lines together are called the coordinates.

To define any point in the space covered by x and y, it will suffice to trace perpendiculars from the point to the cordinates x and y.

2) Mathematical functions

Let ut take two measurements x and y: if y varies in correspondence to the variation of x, we say that y is a function of x and will write:

   y = f(x)

where x is the independent variable and y the dependent variable.

56

3) Graph of directly proportional function

This occurs when y increases with the increase of x and is written:

y = a × x

The graph will be a straight line passing through the origin of the axes (coordinates).

Example: To buy one button (y) I spend one penny (x):

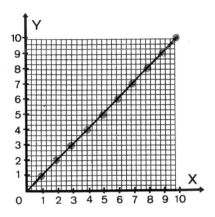

Fig. 46

4) Graph of inversely proportional function

This occurs when y decreases with the increase of x, and is written

$$y = \frac{a}{x}$$

The graph will be an equilateral hyperbole.

Example: To do a certain job a worker (x) takes 10 hours (y):

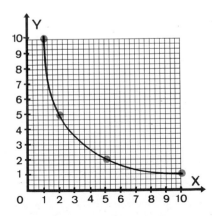

Fig. 47

D - **PROGRESSIONS**

1) **Arithmetic progressions**

This is a series of numbers in which each number is equal to the preceding number plus (+) a given number (called "the common difference").

Examples:

1, 2, 3, 4, 5, 6 . . . . (common difference = + 1)
2, 4, 6, 8, 10, 12 . . . . (common difference = + 2)
10, 20, 30, 40, 50, 60 . . . . (common difference = + 10)

2) **Geometric progressions**

This is a series of numbers in which each number is equal to the preceding number multiplied by (x) a given number (called "a common ratio").

Examples:

10, 100, 1000, 10000 . . . . (common ratio = x 10)
2, 4, 8, 16, 32 . . . . (common ratio = x 2)

E - **MEASUREMENT OF AN ANGLE**

1) **Measure in degrees**

The angle is a part of a plane enclosed between two straight lines having the same origin which is called the apex (vertex).

Fig. 48

When the two lines are perpendicular one with the other the angle is called a *"right angle"* and measures 90°.

Fig. 49

When the angle includes the whole plane, that is, corresponds to 4 right-angles, it is referred to as a *circle angle* and measures (90° x 4) = 360°

Fig. 50

58

The combination of two right-angles gives a *straight angle* of 180°.

Fig. 51

Angles which are greater than a right-angle are called *obtuse* and measure more than 90°: angles less than a right-angle are called *acute* and measure less than 90°

Fig. 52

The measurment of angles is simplified with an instrument known as a "protractor".

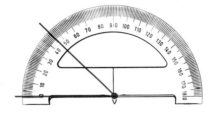

Fig. 53

It suffices to position the base of the protractor over one side of the angle to be measured with the point V over the apex of the angle and then to read the degrees at the point in which the other side intersects the "hemisphere" of the protractor.

2) Measurement in trigonometrical tangents

The size of an angle less than 90° can be calculated also on the basis of two sides of a triangle. We can explain this better with the aid of a diagram - Fig. 54:

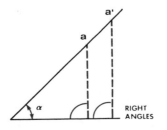

Fig. 54

The angle α (the Greek letter "alpha") can become a right-angle triangle if, from any point (a, a') on one side we construct a line perpendicular to the other side.

We can note that the two triangles formed by a and a' are similar, and, in consequence, the ratio between their component sides will be the same.

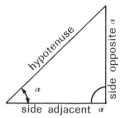

Fig. 55

In our right-angle triangle we shall have a hypotenuse (the side opposite the right-angle) and two shorter sides (adjacent to the right-angle) which, in respect of the acute angle α, we can consider as one being adjacent to α and the other opposite to α.

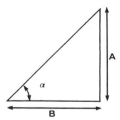

Fig. 56

Well then, the tangent (tan.) of the acute angle α of our right-angle triangle is the ratio between the opposite side to α (A) and the side adjacent to α (B).

$$\tan \alpha = \frac{A}{B}$$

The tangent of an angle α of 0° will equal 0
  »   »   » »   »   » » 45° will equal 1
  »   »   » »   »   » » 90° will equal ∞ (infinity)

The following table gives the tangents of angles from 1° to 90°.

TABLE OF TANGENTS OF ANGLES FROM 1° TO 90°

Angle in degrees	Tangent	Angle in degrees	Tangent	Angle in degrees	Tangent
1°	0.01	31°	0.60	61°	1.80
2°	0.03	32°	0.62	62°	1.88
3°	0.05	33°	0.65	63°	1.96
4°	0.07	34°	0.67	64°	2.05
5°	0.09	35°	0.70	65°	2.14
6°	0.11	36°	0.73	66°	2.25
7°	0.12	37°	0.75	67°	2.36
8°	0.14	38°	0.78	68°	2.48
9°	0.16	39°	0.81	69°	2.61
10°	0.18	40°	0.84	70°	2.75
11°	0.19	41°	0.87	71°	2.90
12°	0.21	42°	0.90	72°	3.08
13°	0.23	43°	0.93	73°	3.27
14°	0.25	44°	0.97	74°	3.49
15°	0.27	45°	1.00	75°	3.73
16°	0.29	46°	1.03	76°	4.01
17°	0.31	47°	1.07	77°	4.33
18°	0.32	48°	1.11	78°	4.70
19°	0.34	49°	1.15	79°	5.14
20°	0.36	50°	1.19	80°	5.67
21°	0.38	51°	1.23	81°	6.31
22°	0.40	52°	1.28	82°	7.12
23°	0.42	53°	1.33	83°	8.14
24°	0.45	54°	1.38	84°	9.51
25°	0.47	55°	1.43	85°	11.43
26°	0.49	56°	1.48	86°	14.30
27°	0.51	57°	1.54	87°	19.08
28°	0.53	58°	1.60	88°	28.64
29°	0.55	59°	1.66	89°	57.29
30°	0.58	60°	1.73	90°	∞

DENSITOMETRY

Densitometry, as we have already said, is the technique of measuring the *blackening* acquired by a photographic emulsion which has been exposed, developed and fixed.

The blackening (which, as we know, is due to metallic silver) can be more or less strong or, as is more customary to say in photography, more or less dense. Thus the terms "blackening" and "density" have the same significance in photography.

The blackening (or photographic density) depends principally on three elements: 1) exposure to light, 2) type of emulsion, 3) kind of developer.

Hence, for a given type of emulsion and given kind of developer, it will be sufficient to graduate and control the exposure to light so as then to be able to measure the corresponding densities acquired from the emulsion. This graduation of light is affected by means of the "grey scale" or step wedge.

1) THE GREY SCALE

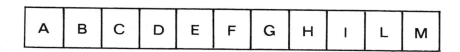

Fig. 57

Suppose we wish to illuminate various "steps" in a strip of film (see Fig. 57) with progressively decreasing exposures, on a ratio of x 0.5 (halving), so that the exposure received by step B is half that received by step A ($E_B = \dfrac{E_A}{2}$), the exposure received by C is half that received by B ($E_C = \dfrac{E_B}{2}$), and so on.

Fig. 58

The most easily understandable method to obtain this result is to superimpose one over the other so many small strips of grey glass, each absorbing half the light falling on it, of diminishing lengths to as to form a scale of 10 steps, as shown in Fig. 58.

By putting this scale of greys over the strip of photographic film, we can observe that every step along the film (with the exception of A which remains uncovered) will have an increasing number of layers of grey glass from 1 (step B) to 10 (step M).

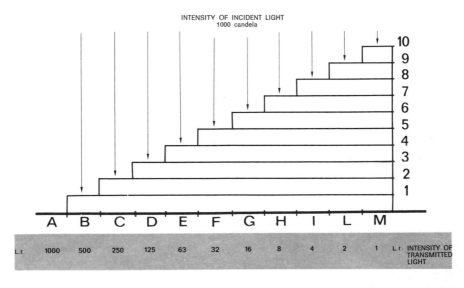

Fig. 59

If we place above the "grey scale" (shown schematically) a light source having an intensity of 1000 candelas (incident light = Li), the light passing through the "grey scale" (transmitted light = Lt) in the various steps will be:

Step A	Lt =	candela	1000/1	= 1000	candelas
Step B	Lt =	»	1000/2	= 500	»
Step C	Lt =	»	$1000/2^2$	= 250	»
Step D	Lt =	»	$1000/2^3$	= 125	»
and so on until					
Step M	Lt =	»	$1000/2^{10}$	= 1	»

With this preamble, we can now introduce the two ideas of Transmission and Opacity, and then define what interests us most, that is, the protographic density.

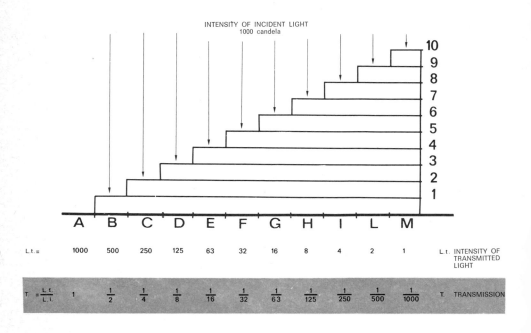

Fig. 60

a) The light-passing capacity of a grey scale or more to the point, a photographic image, is called its *Transmission* (T) and is the ratio between the intensity of the light transmitted (Lt) and the incident light (Li).

$$1) \quad T = \frac{Lt}{Li}$$

Thus in the case of our grey scale we have:

64

Step A $\qquad$ $T = \dfrac{1000}{1000} = 1$

Step B $\qquad$ $T = \dfrac{500}{1000} = \dfrac{1}{2}$

Step C $\qquad$ $T = \dfrac{250}{1000} = \dfrac{1}{4}$

Step D $\qquad$ $T = \dfrac{125}{1000} = \dfrac{1}{8}$

and so on until

Step M $\qquad$ $T = \dfrac{1}{1000} = \dfrac{1}{1000}$

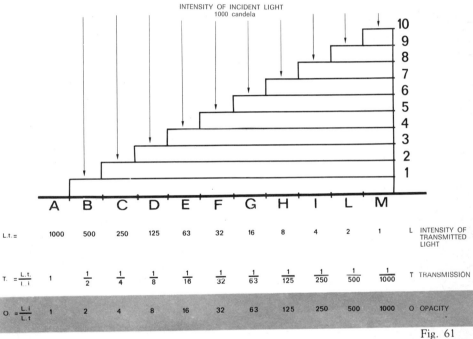

Fig. 61

b) The *Opacity* (O) of a photographic image is the ratio between the intensity of the incident light (Li) and the transmitted light (Lt), that is, the inverse of the Transmission:

$$2) \quad O = \frac{Li}{Lt} = \frac{1}{T}$$

Hence in our grey scale we shall have:

Step A $O = \dfrac{1000}{1000}$ $= 1$

Step B $O = \dfrac{1000}{500}$ $= 2$

Step C $O = \dfrac{1000}{250}$ $= 4$

Step D $O = \dfrac{1000}{125}$ $= 8$

and so on until

Step M $O = \dfrac{1000}{1}$ $= 1000$

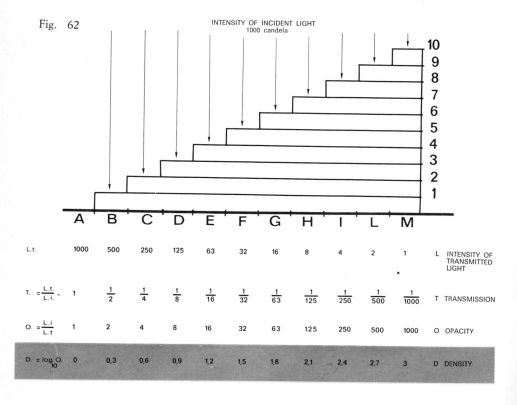

Fig. 62

INTENSITY OF INCIDENT LIGHT
1000 candela

	A	B	C	D	E	F	G	H	I	L	M	
L.t.	1000	500	250	125	63	32	16	8	4	2	1	L INTENSITY OF TRANSMITTED LIGHT
$T. = \dfrac{L.t.}{L.i.}$	1	$\dfrac{1}{2}$	$\dfrac{1}{4}$	$\dfrac{1}{8}$	$\dfrac{1}{16}$	$\dfrac{1}{32}$	$\dfrac{1}{63}$	$\dfrac{1}{125}$	$\dfrac{1}{250}$	$\dfrac{1}{500}$	$\dfrac{1}{1000}$	T TRANSMISSION
$O. = \dfrac{L.i}{L.t}$	1	2	4	8	16	32	63	125	250	500	1000	O OPACITY
$D. = \log_{10} O.$	0	0,3	0,6	0,9	1,2	1,5	1,8	2,1	2,4	2,7	3	D DENSITY

c) The logarithm of the Opacity (O) is called the photographic *Density:*

3) $D = \log_{10} O = \log_{10} \dfrac{1}{T}$

66

By referring to the log. table on page 55, we can find the Density values of our "grey scale":

Step A   D = log 1 = 0
Step B   D = log 2 = 0.3
Step C   D = log 4 = 0.6
Step D   D = log 8 = 0.9
and so on until
Step M   D = log 1000 = 3

In a screen (dot) image, the relationship between its mean density and *the percentage of the dot area* is given in the following scale:

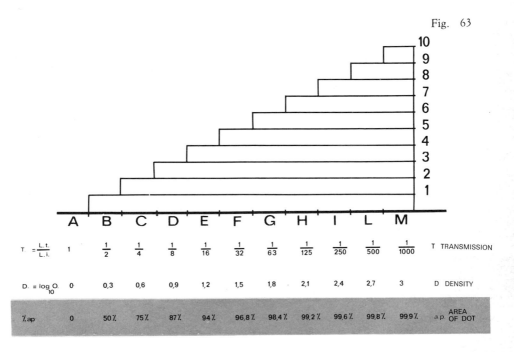

Fig. 63

In fact for a Transmission of 1 (that is, a percentage transmission of 100%) the Density = 0.0 corresponds to a dot area of 0%.

Footnote (1) See page 54 for the nature of logarithms.

Here one might ask, why complicate things by introducing logarithms when Density and Opacity are no more than two different ways of saying the same thing?

However, the use of the logarithm of the Opacity — that is, the Density — is in fact motivated by big advantages, the chief of which are as follows:

For a Transmission of ½ (50%) D = 0.3 corresponds to a dot area of 50%.

For a Transmission of ¼ (25%) D = 0.6 corresponds to a dot area of 75%.

For a Transmission of $\dfrac{1}{1000}$ (0.1%) D = 3 corresponds to a dot area of 99.9%.

d) The logarithm of the relative Exposure (log. E rel.)

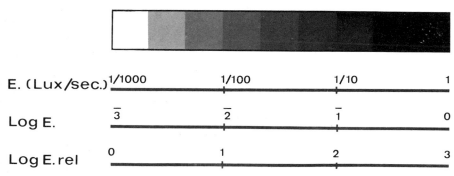

Fig. 64

What chiefly interests us in studying the characteristics of a photographic emulsion is — as already noted — to subject the emulsion to an *progressively increasing series of exposures* through a grey scale of known Density range.

---

1) The numbers applying to Density are much smaller than those applying to Opacity, for example: 1.5 instead of 32, 3 instead of 1000.

2) The Density scale increases with an arithmetic progression while that for Opacity with a geometric progression.

3) Calculations are greatly simplified by having additions and subtractions instead of multiplications and divisions.

4) Above all, the numerical increases in Density correspond to the *apparent* increases in density.

For example, in superimposing *two* films of D = 1.2 we obtain a D = 2.4 which is exactly 1.2 + 1.2. Expressed in opacity, the *two* films having O = 16 will have a combined opacity of 256, a figure which suggests that a great many film have been superimposed...

For the same reasons, it is customary to use logarithmic numbers to express the Exposure (log. E) instead of arithmetic values.

After the film has been processed, the densities corresponding to the scale of exposures are read with a densitometer, thus giving us two series of data (the Exposure and the corresponding Density) which must now be related.

Since the Densities are expressed in logarithmic numbers, the relationship can be facilitated by applying logarithmic numbers also to the Exposure scale. (See also Footnote on page 67-68).

Finally, since the Exposure needed by modern photographic emulsions is very small (of the order of tenths of lux-sec) thereby coming into the region of negative logarithms, it is convenient to use, not the logarithms of the actually exposure applied (log. E), but the logarithms of the Exposures "relative" to those effectively used (log. E. rel.), even though they are greater.

In this way (without changing the ratios between the various exposures) we obtain logarithms of positive characteristic (from 0 to 3) as for Density, which greatly facilitates all the calculations we have to make.

Note: The "photographic grey scale" and "photographic step-wedge".

In practice one uses a "photographic grey scale" which is a strip of film having increasing Densities, which correspond to the solid scale.

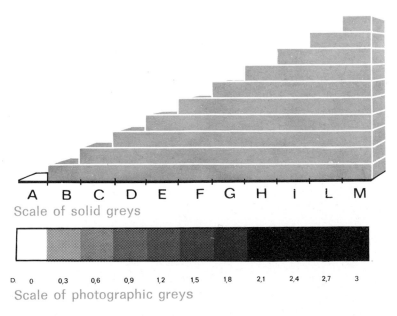

Fig. 65

The "photographic grey scale" has three basic characteristics: the number of steps, the increment of D and the range of D.

In the example we have been using, the photographic step-wedge has the following characteristics:

Number of steps = 11
Density increment per step = 0.3
Density scale = 0 to 3.0

For more precise measurements, scales of 20 or 30 steps with density increments of 0.15 or 0.1 are used.

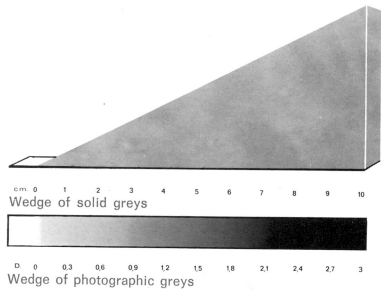

cm. 0   1   2   3   4   5   6   7   8   9   10
Wedge of solid greys

D. 0   0,3   0,6   0,9   1,2   1,5   1,8   2,1   2,4   2,7   3
Wedge of photographic greys

Fig. 66

On occasions, a scale of continuous greys is used instead of one in steps when high precision is needed. It may have the same density range as one in steps, but the increment in density is continuous. In this case the characteristics may be specified as follows:

Length of the wedge = 11 cm.
Density increment per cm. = 0.3
Density scale = 0 to 3.0

Photographic wedges, whether in steps or continuous, are available both on photographic film for the measurement of original transparencies and on paper for the measurement of opaque originals photographed by reflected light.

e) Contrast and density scale

Fig. 67

The Contrast is the difference of blackening which we encounter comparing any two tones of an image, for example A and B.

Expressed in arithmetic values (Opacity) the contrast between A and B will be $\dfrac{A}{B}$. Expressed in logarithmic values (Density) the contrast between A and B will be A — B.

Fig. 68

On the other hand, *the Density scale* is the difference we encounter between *the two extreme densities* of the image - the blackest, called Maximum (M) and the most transparent, called the minimum (m) density of the image, for example, A_1 and B_1.

In terms of the Opacity, the scale between A_1 and B_1 will be $\dfrac{A_1}{B_1}$ or in terms of Density, the scale between A_1 and B_1 will be $A_1 — B_1$.

2) THE CHARACTERISTIC CURVE

In any original (subjects on glass, drawings or paintings, photographic positives or negatives, etc.) the various Densities are distributed at random, thus making it extremely difficult to make a systematic check of the photographic reproduction.

By placing a grey scale at the side of an original (suppose we take a continuous tone negative), we shall have the various Densities of the original well arranged according to a Density increment known to us (Fig. 69 a).

The "grey scale" is in fact like a summary of all the tones of the original.

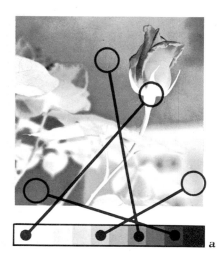 

a   b

Fig. 69

After reproduction (Fig. 69 b), we can check the accuracy of the reproduction (without concerning ourselves with the subject photographed) solely by comparing the two grey scales, that of the original and that reproduced.

This comparison is made by means of the "characteristic curve".

Let us see how it is constructed and how it is interpreted.

A) How the curve is constructed

The procedure is as follows:

1) Taking a strip of the photographic film to be tested, a known series of exposures are made by means of a grey scale.

72

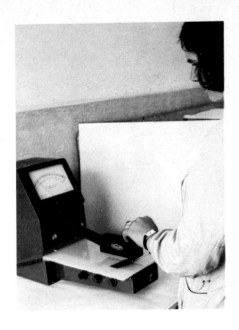

Fig. 70

2) The strip is processed and the Densities so obtained measured with a densitometer. Suppose we obtain the following results:

Grey scale - original

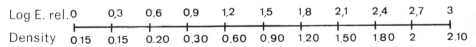

Log E. rel.	0	0,3	0,6	0,9	1,2	1,5	1,8	2,1	2,4	2,7	3
Density	0.15	0.15	0.20	0.30	0.60	0.90	1,20	1,50	1,80	2	2.10

Grey scale - reproduction

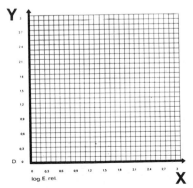

Fig. 71

73

3) The two coordinates are drawn on a sheet of graph paper and are calibrated with the values of the relative log. E (independent variables) along the horizontal axis (x) over an range 0 to 3.0, and along the vertical axis (y) with a Density range from 0 to 3.0 - see Fig. 71.

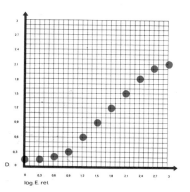

Fig. 72

4) The Densities obtained from the test strip are now plotted as a series of points on the graph paper (Fig. 72).

Each point thus represents a pair of values Exposure-Density, which will tell us what density corresponds to a certain exposure.

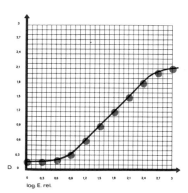

Fig. 73

5) By joining up the various points we obtain a "curve" which shows us the characteristics of the emulsion with a given development. For this reason the curve is called the "characteristic curve of the emulsion and development" or more simply, the "characteristic curve". (Fig. 73).

1) *The four regions of the curve*

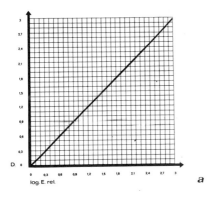
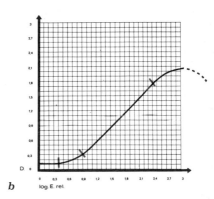

Fig. 74

Theoretically, the Densities of a perfect photographic reproduction should be directly proportional to the Exposure received, giving in graphic form a straight line originating from the point of intersection of the axes (see page 57) as shown in Fig. 74 *a.*

In reality the emulsion shows a certain inertia to the action of light, after which there is usually a region of perfect correspondence between exposure and density, followed by a progressive loss of sensitivity until no further action takes place. See Footnote [1]) This gives a curve in the form of an elongated S with which we are already familiar (Fig. 75).

In this elongated S, we can usually distinguish three regions: the initial curved line, called the "toe" or "foot"; the straight line; and the final curve called the "shoulder".

A more precise and useful subdivision, distinguishes four regions: the region of fog; the region of under-exposure; the region of correct exposure; and the region of over-exposure (see Fig. 74 *b*).

Footnote (1) We might compare it with an automobile which has to travel for 1 mile from a stationary position. Because of the inertia that it must overcome, it will increase its speed progressively over the first few yards until it reaches a steady speed which can be maintained in terms of distance and time. Towards the end of the track, before it can stop, it must diminish its speed progressively.

Let us examine these four regions separately (Fig. 75):

a) *The fog.* This is shown by a small horizontal line at the beginning of the curve. This tells us that the film will possess a minimum density even when it has received no exposure which is due both to the actual density of the base (not perfectly transparent) and to the phenomenon of physical and chemical fog (see page 33) which occurs during development.

b) *Region of under-exposure.* This is the curved line of the "toe" which extends from 0.5 to 0.9 on the relative log. E in Fig. 75. In this region the increase in Density *does not keep pace* with the increase of **Exposure** resulting in an imperfect rendering of the original. From Fig. 75 it can be seen that an Exposure increase from 0.5 to 0.8 gives the same Density increase of 0.1 as an increase in Exposure from 0.8 to 0.9 which is only a third as great.

c) *Region of correct exposure.* This is the straight line which extends in the graph from log. E. rel. 0.9 to 2.4. In this region there is a constant increase in Density with *a corresponding increase in Exposure,* thus giving a perfect rendering of the original (proportional increase).

— In our case the rendering is perfect because with each increase of 0.3 in log E. rel., there is a corresponding increase of 0.3 in Density. The slope of the line is 45°.

— If, with every 0.3 increment in Exposure the increase in Density were 0.4, the rendering would still be proportional, but not identical, in being more contrasty (1). In this case the slope of the straight line would be steeper, that is 60°.

— If, with each increase of 0.3 in Exposure, the corresponding increase in Density were 0.2, the rendering would be proportional, but less contrasty. In this case the slope of the straight line is less, that is 30°.

d) *Region of over-exposure.* This is the curve of the shoulder which extends from log. E. rel. 2.4 to 3.

In this region also (as in that of under-exposure) the increase in Density *does not correspond* to the increase in Exposure with consequent imperfect rendering of the original (non-proportional increase).

Footnote (1) We shall deal later with the idea of « contrast » of an emulsion (page 78). For the present it is enough to know that it is shown by the « slope » of the characteristic curve.

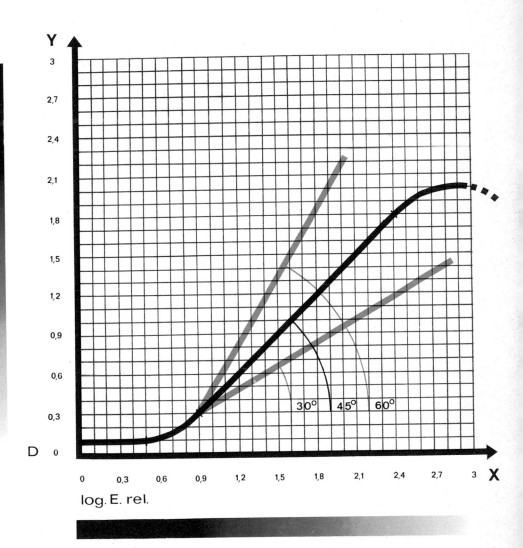

Fig. 75

It is interesting to note that if we continue to increase the exposure there can be a *decrease* in Density (so-called solarisation) which would be shown by the broken line in Fig. 75.

This phenomenon has been put to use in special emulsions designed to give positive images directly (autopositives).

2) *The Slope or Gradient.* A further step in the interpretation of the characteristic curve is the study of its "slope", that is, its gradient.

This, being a slope, is measured by the size of the angle which the extension of the curve forms with the horizontal axis. In sensitometry, the amplitude of this angle is expressed by means of its tangent (see page 59).

We can distinguish between three types of gradient: the gradient, the mean gradient, and the maximum gradient.

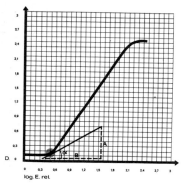

Fig. 76

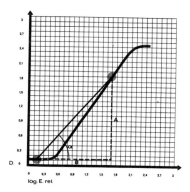

Fig. 77

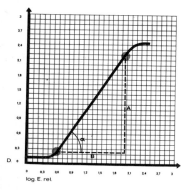

Fig. 78

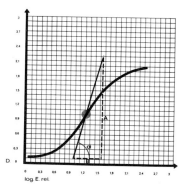

Fig. 79

a) *The Gradient* (G). It is the slope of a *non-rectilinear* part of the curve at *a point* on it (Fig. 76). It corresponds to the tangent of the curve at that point, that is, to the tangent of the angle α (footnote 1).

$$G = \frac{A}{B} = \tan. \alpha$$

b) *The mean Gradient* ($\overline{G}$). It is the slope of the straight line that joins *two points* of a *non-rectilinear* part of the curve. It is the mean of all the gradients of the curve and is expressed by the tangent of the angle α (Fig. 7).

$$\overline{G} = \frac{A}{B} = \tan. \alpha$$

When the mean gradient of the *whole curve* (that is, the slope of the line which joins the extreme points of the curve), we obtain a general contrast indication or "gradation" of the emulsion.

c) *The maximum Gradient or Gamma* (γ). This is the slope of the *straight line region* of the characteristic curve (Fig. 78). It is therefore given as the tangent of the angle formed by the extension of the straight line with the horizontal axis.

$$\gamma = \frac{A}{B} = \tan. \alpha$$

When there is no straight line region in the characteristic curve (Fig. 79) the gamma is the slope ($\frac{A}{B}$) of the curve *at the point* where the regions of increasing and decreasing density meet.

Let us now examine further a most important aspect of gamma:

Gamma (γ) *is the "contrast factor" of an emulsion.*

Every photographic emulsion has its particular capacity for rendering the contrast of the original. From this point of view, it is usual to classify photographic emulsions into those of "normal" gradation, those of "steep" gradation (high contrast) and those of "soft" gradation. The curves which follow illustrate this characteristic:

Footnote (1) The tangents of angles from 0 to 90° range from 0 to ∞ (see page 60).

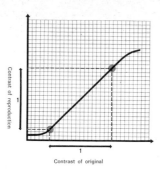

Fig. 80

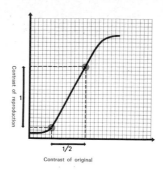

Fig. 81

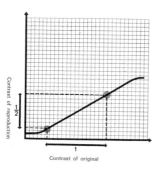

Fig. 82

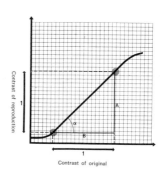

Fig. 83

A photographic emulsion of normal gradation (one that gives a photographic reproduction with contrast equal to that of the original, Fig. 80), will have a characteristic curve with a gradient of 45° that is $\gamma = 1$.

A "contrasty" emulsion (one with a contrast greater than that of the original, Fig. 81), will have a characteristic curve with gradient $> 45°$ (greater than 45°), that is $\gamma > 1$.

An emulsion of soft gradation (one that has a contrast lower than that of the original, Fig. 82), will have a characteristic curve with gradient $< 45°$ (less than 45°), that is, with $\gamma < 1$.

The different "contrast renderings" of the three emulsions are thus clearly shown by the slope of the characteristic curve which, as we know, is measured by the ratio $\dfrac{A}{B}$ (Fig. 83), which, in turn, is the tangent of the angle α, that is, the gamma γ. For this reason gamma is called the "contrast factor" of the emulsion.

The equation of gamma

We are now in a position to state the mathematic formula which governs relationship between the three sensitometric quantities, namely, the Exposure (log. E. rel.), the Density (D) and the gamma (γ).

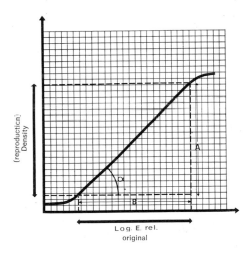

Fig. 84

By definition the value of γ is given by the tangent of the angle α, that is $\dfrac{A}{B}$

$$\gamma = \frac{A}{B}$$

and since A represents the Density of the reproduction, while B represents the logarithm of the relative Exposure, it follows that:

$$1)\quad \gamma = \frac{D}{\log.E.rel.}$$

From this it is now easy to deduce the value of the Density:

$$2)\quad D = \gamma \times \log.E.rel$$

and hence the value of the relative log. Exposure:

$$3)\quad \log.E.rel. = \frac{D}{\gamma}$$

81

Extension of the "equation of gamma"

Looking carefully at the values: γ, log. E. rel., and D of the equation of gamma, we can note that:

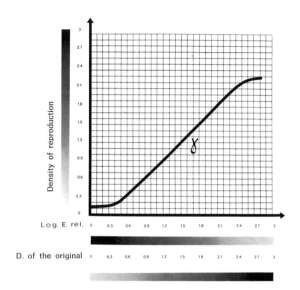

Fig. 85

1) the γ is obviously the γ of the reproduction.

That D is also obviously the D of the reproduction.

The exposures (log. E. rel.) correspond to the Densities of the original: in fact *both increase in value* from 0 to 3 (Fig. 85).

Hence, we can always substitute for the values of log. E. rel., the Density values of the original.

We shall then have:

$$1) \ \gamma \text{ of the reproduction} \ = \ \frac{\text{D of the reproduction}}{\text{D of the original}}$$

from this:

$$2) \ \text{Density of reproduction} = \gamma \text{ of reproduction x D of original}$$

and

$$3) \ \text{Density of original} \ = \ \frac{\text{D of the reproduction}}{\gamma \text{ of reproduction}}$$

2) The Density of the original and that of the reproduction can be considered in general (as in the "equation of gamma") or from particular aspects:

as "Minimum Density" (e.g., Dm = 0.3)
as "Maximum Density" (e.g., DM = 1.8)
as "Scale of Density" or Density Range (e.g., Sc.D = 1.5)
as "Increase or decrease of Density" (e. g., Density increase 0.2).

We shall therefore have from 1):

$$\gamma \text{ of reproduction} = \frac{\text{Scale of D. of reproduction}}{\text{Scale of D. of original}}$$

or, from 3):

$$\text{Dm of original} = \frac{\text{Dm of reproduction}}{\gamma \text{ of reproduction}}$$

and so on.

3) FROM THEORY TO PRACTICE

The practical objective that we wish to achieve is to obtain photographic reproductions from any original:

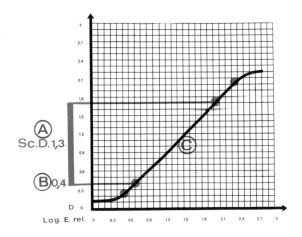

Fig. 86

A) with a predetermined Scale of Density, (for example, 1.3)
B) with a predetermined minimum Density (and hence, assuming the Density scale is constant, of known Maximum D) - for example, Dm = 0.4.
C) with the whole Scale of Densities falling in the region of correct exposure.

A) How to obtain the required Scale of Densities

Principle:

The Scale of Densities depends on the gamma.

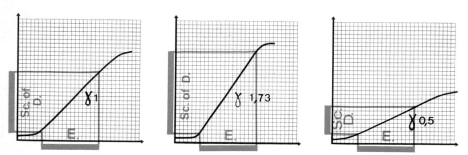

Fig. 87

In fact (Fig. 87), given equal Exposure, the Scale of Densities varies as the gamma varies. The problem is thus one of controlling the gamma. Let us see how:

Above all, the conditions of development must be standardised, that is, the *temperature* must be maintained at 20° C and the rate of *agitation constant.* While it is true that an increase of temperature and agitation have the effect of increasing the gamma and hence the contrast, their effect on the gamma is difficult to predict and it is thus best to disregard this means of controlling the gamma. *By far the best procedure is standardisation.*

Having said this, we have three ways of influencing the gamma:

1) Above all by choosing an emulsion of suitable gradation, that is, with a "contrast factor" which closely approaches what we require (see page 79).

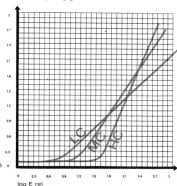

Fig. 88

In Fig. 88 we show the characteristic curves of three continuous tone (Graphic Tone) emulsions of Ferrania-3M of soft gradation (LC), normal

(MC) and contrasty (HC), developed for the same time in the same developer (Graphic Powder) at a temperature of 20° C with continuous agitation.
Their gamma can give us an initial guide in the choice of an emulsion.

2) A further correction of the gamma is open to us in using a suitable
developer, that is one giving greater or less contrast (see page 34).

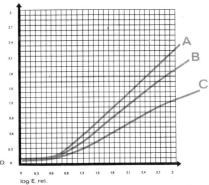

Fig. 89

In Fig. 89 we show the characteristic curves of the same emulsion
(Graphic Tone LC) developed for the same time (4 minutes), temperature
(20°) and agitation (continuous) in three different developers: A, high
contrast (Graphic Powder); B, normal contrast (Graphic Roll); C, low
contrast (Delofin).

3) But once the choice of the most suitable emulsion and developer has
been made, the *determining factor* on gamma will be that of the *duration*
(time) of development and the temperature.
This time control is very important both in terms of the wide range of
control its gives over gamma and for the accuracy of the results which can be
obtained from it.

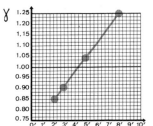

Time of Development (min.)

Fig. 90

In this respect "time-gamma" curves which show what the gamma will
be for a given time of development are extremely useful.

85

In Fig. 90 we show the curve of Graphic Tone LC film developed in Graphic Powder developer at 20° C with continuous agitation:

From it we can predetermine that to obtain, for example, a gamma of 0.85 we must develop for 2 minutes; if we want to obtain a gamma of 0.90 we must develop for 3 minutes; if we want a gamma of 1.03 we must develop for 5 minutes; if we want a gamma of 1.25 we must develop for 8 minutes.

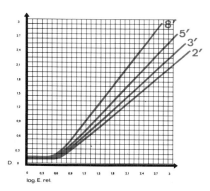

Fig. 91

The characteristic curves so obtained enable us to assess the effect on gamma of the time of development.

B) How to obtain a required minimum Density

Principle:

For the same gamma the minimum Density depends on the exposure.

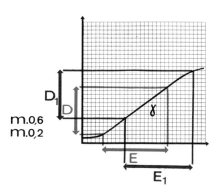

Fig. 92

In fact by increasing the Exposure from E to E_1, the Densities, while still retaining the same scale (range) increase the minimum from 0.2 to 0.6.

C) **How to obtain a correct scale of Densities**

Principle:

   In increasing or decreasing the Exposure the "scale of correct exposures" must not be exceeded.

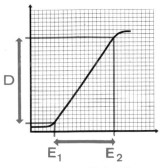

Fig. 93

   In fact the scale of Densities will be "correct" only if the Exposures necessary to obtain it are within the projection of the straight line region of the curve on to Exposure axis itself. Such projection (Fig. 93, E_2 - E_1) is called the "scale of correct exposures" or simply, "exposure scale".
   It can be seen that the possibility of increasing or decreasing the Exposure will be given by the difference (logarithmic) between the "scale of correct exposures" and the Density scale of the original. Such a difference is called the *"Exposure Latitude"*.

Exposure Latitude = Exposure scale — Density scale of original

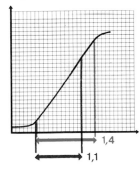

Fig. 94

   Thus, if a given emulsion possesses an Exposure Scale of 1.4, and the Density Scale of an original is 1.1, the exposure latitude will be 1.4 — 1.1 = 0.3 (antilog = 2).
   This means that the minimum exposure sufficient to reproduce all the Densities of the original can be increased up to 2 times.

N.B.

In practice the exposure may sometimes be such as to exceed the straight line region of the curve so as to make partial use of the toe or shoulder region. In this case the Exposure Scale becomes greater, for example, from 1.4 to 1.7 as shown in Fig. 95 and hence the effective Exposure Latitude becomes greater.

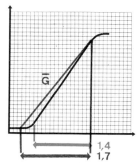

Fig. 95

Such an advantage, however, will be accompanied by two drawbacks: the lowering of gamma that is substituted by the mean gradient G, and the imperfect (it would be better to say "non-proportional") reproduction of the densities of the original.

D) **Practical examples**

To assist in the understanding and application of what we have so far discussed, let us illustrate the two cases of sensitometric calculations which apply most to practice.

The first concerns the modification of the Scale of Densities, and the second the Scale of Densities and at the same time the minumum Density.

Remember that the easiest method of achieving certain results is that of making initial tests in one's own laboratory and from these to plot the " characteristic curves" and "time-gamma" curve of the emulsion to be used.

Once this has been done, it only remains to apply the "equation of gamma" which we have already referred to (see pages 81-83), for whatever result we desire to obtain.

The preparation of "characteristic" and "time-gamma" curves

The first step is to cut five strips from a sheet of the film to be tested and to give each of them the same basic exposure through a photographic step wedge (grey scale).

The exposure given can be based on previous experience of similar films or found by means of tests. Very often the technical literature accompanying the film will give a guide.

The strips are processed in the same developer, the first being withdrawn after 2 minutes (and placed in the stop bath), and with subsequent strips at 3, 4, 5 and 6 minutes. They are then fixed, washed and dried.

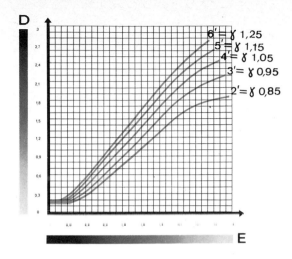

Fig. 96

The densities of the strips are then measured with a densitometer and from these are plotted the five characteristic curves in the way already explained on page 72.

The angles of the gradients are then measured with a protractor and by using the table on page 61, the corresponding values of the tangents are written at the side of each curve as shown in Fig. 96.

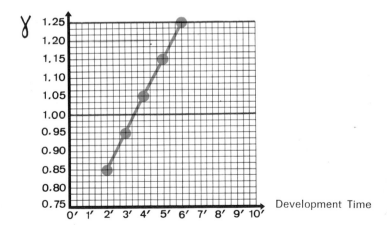

Fig. 97

When this has been done, the "time-gamma" curve can now be plotted using the horizontal axis for the time of development in minutes and the vertical axis for the values of gamma, as shown in Fig. 97.

For the convenience of the reader we repeat the "equation of gamma" already explained on pages 81-83 with the derivations from it:

$$1)\quad \gamma = \frac{D}{\log E \ rel}$$

that is:

$$\text{gamma of reproduction} = \frac{\text{Density of reproduction}}{\text{Density of original}}$$

from which we deduce:

$$2)\quad D = \gamma \times \log E \ rel$$

that is:

Density of reproduction = γ reproduction x Density of original

and then:

$$3)\quad \log E \ rel = \frac{D}{\gamma}$$

that is:

$$\text{Densities of original} = \frac{D \text{ of reproduction}}{\gamma \text{ of reproduction}}$$

Let us now look at the first of our two examples.

1st CASE: Modification of the Scale of Densities

We know already that the Scale of Densities depends on gamma and that the gamma is controlled by the time of development.

The following three examples illustrate the calculation for obtaining a reproduction with Scale of Densities greater, less and equal to that of the original.

EXAMPLE No 1: We wish to obtain a reproduction with Scales of Densities greater than the original.

Let us suppose we have an original positive with a Scale of Densities of 1.30 and we wish to obtain a negative reproduction with a Scale of Densities of 1.50:

Applying the equation of gamma we have:

$$\gamma \text{ of reproduction} = \frac{\text{Sc. D of reproduction}}{\text{Sc. D of original}} = \frac{1.50}{1.30} = 1.15$$

Referring now to the "time-gamma" curve, we shall see that to obtain a gamma of 1.15 (to which a Sc of D of 1.50 corresponds) we must develop the film for 5 minutes.

EXAMPLE No 2: We wish to obtain a Scale of Densities less than the original.

If the original had had a Scale of D of 1.60, the time of development

90

would have been:

$$\gamma = \frac{1,50}{1,60} = 0,95$$

which corresponds to 3 minutes on the "time-gamma" curve.

EXAMPLE No 3: *We wish to obtain a reproduction with Scale of Densities equal to the original.*
 If, from the same original (Sc. D = 1.60) we wish to obtain a reproduction having the same Sc. D, the development time would have to be:

$$\gamma = \frac{1,60}{1,60} = 1$$

which corresponds to a time of 3.6 minutes in the "time-gamma" curve.

2nd CASE: Modification of both the Scale of Densities and the minimum D.

 We know that the modification of the Scale of D is obtained by modifying the *time of development,* while modification of the minimum D. is obtained by modifying the *exposure.*
 The four examples which follow illustrate how we can obtain a reproduction with Sc. D and Dm greater than the original: with Sc. D and Dm less than the original; and with Sc. D *less* and Dm *greater* than the original.

EXAMPLE No 1: *We wish to obtain a reproduction with Sc. D and Dm greater than the original.*
 From an original with Sc. D = 1.30 and Dm = 0.20 we wish to obtain a reproduction with Sc. D = 1.50 and Dm = 0.35.

Original		Reproduction wanted	
D Max.	= 1.50	D Max	= 1.85
D min.	= 0.20	D. min.	= 0.35
Sc. D	= 1.30	Sc. D	= 1.50

 The time of development will be modified on the basis of gamma, applying formula 1):

$$\gamma = \frac{\text{Sc. D reproduction}}{\text{Sc. D original}}$$

$$\gamma = \frac{1.50}{1.30} = 1.15$$

which corresponds to a time of 5 minutes on the time-gamma curve.
 The basic Exposure will instead be modified (in this case "increased") in proportion to the difference between the two minimum Densities, that of the original and that of the reproduction, applying formula 3):

$$\log.E.rel. = \frac{D}{\gamma} = \frac{Dm - Dm}{\gamma}$$

$$\log E. \, rel. = \frac{0.35 - 0.20}{1.15} = \frac{0.15}{1.15} = 0.13 \, (antilog = 1.34)$$

The basic exposure must therefore be increased by multiplying it by 1.34.

NB. From the "characteristic curve" of a gamma of 1.15 we can check that the whole Scale of Densities lies on the straight line of the curve, that is, from 0.35 to 1.85. This can, in fact, be checked beforehand.

EXAMPLE No 2: We wish to obtain a reproduction with Sc. D and Dm less than the original.

Original		Reproduction wanted	
D Max.	= 1.80	D Max. = 150	
D min.	= 0.40	D min. = 0.30	
Sc. D	= 1.40	Sc. D = 1.20	

The time of development will be given from:

$$\gamma = \frac{1.20}{1.40} = 0.85$$

which corresponds to 2 minutes on the time-gamma curve.

The Exposure will be given by the basic Exposure decreased by:

$$\log E. \, rel. = \frac{0.40 - 0.30}{0.85} = \frac{0.10}{0.85} = 0.12 \, (antilog = 1.32)$$

The basic exposure must therefore be decreased and is therefore divided by 1.32.

EXAMPLE No 3: We wish to obtain a reproduction with Sc. D greater and Dm less than that of the original.

Original		Reproduction wanted	
D Max.	= 1.20	D Max. = 1.30	
D min.	= 0.40	D min. = 0.30	
Sc. D.	= 0.80	Sc. D = 1	

The time of development will be given by:

$$\gamma = \frac{1}{0.80} = 1.25$$

which corresponds to 6 minutes in the time-gamma curve.

The Exposure will be given by the basic Exposure decreased by:

$$\log.E.rel. = \frac{0.40 - 0.30}{1.25} = \frac{0.10}{1.25} = 0.08 \, (anti \, log. = 1.2)$$

The basic exposure is therefore divided by 1.2.

EXAMPLE No 4: We wish to obtain a reproduction with Sc. D less and Dm greater than the original.

Original
D Max. = 1.00
D min. = 0.10
Sc. D   =0.90

Reproduction wanted
D Max. = 1.15
D min. = 0.30
Sc. D   = 0.85

The time of development will be given by:

$$\gamma = \frac{0.85}{0.90} = 0.95$$

which corresponds to 3 minutes on the time-gamma curve.

The Exposure will be given from the basic Exposure increased by:

$$\log.E.rel. = \frac{0.30 - 0.10}{0.95} = \frac{0.20}{0.95} = 0.21 \text{ (anti log. } = 1.62)$$

The basic Exposure must therefore be multipled by 1.62.

PHOTOGRAPHY IN THE VARIOUS METHODS
OF PRINTING

Subjects which are reproduced by printing methods are of two types: *line* (letters, drawings in ink and in general those subjects in which the design is all of the same density), *and continuous tone* (photographs, pictures, etc., and in general those subjects in which the design is a gradation of different densities).

Fig. 98   Line   Continuous tone

The reproduction by printing of these subjects is done by depositing an extremely thin layer of ink on the paper in such a way as to reproduce the original — whether line or continuous tone — as accurately as possible.

To do this various printing methods are used (photogravure, offset lithography, letterpress and electrostatic printing), each of which require a different photographic preparation of the original.

We shall deal with them separately:

1) PHOTOGRAVURE

A) Method of printing

The printing surface (matrix) is a sheet (or cylinder) of copper on which have been etched minute cells (Fig. 99) having the same area but not the same depth (Footnote 1).

Footnote (1) We are referring to the traditional photogravure which is still widely in use. New techniques of gravure make use of cells of the same depth but of different area (autotype gravure), or cells having different depth and area (semi-autotype gravure). See page 97.

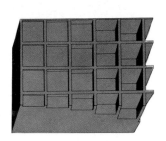

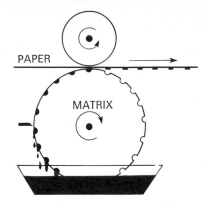

Fig. 99   Fig. 100

In the printing cycle the ink is deposited in the cells from the ink trough and cleaning blade and is then absorbed by the paper (Fig. 100).

The different printing densities are proportional to the different depths of the cells.

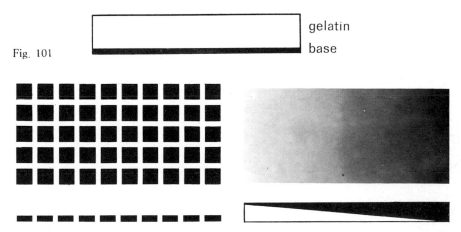

Fig. 101

Fig. 102   Fig. 103

The incision of the cells on the cylinder of copper is obtained as follows: a special "pigment paper" sensitised with potassium bichromate (Fig. 101) is successively exposed through a black screen consisting of small squares separated by clear lines (Fig. 102); the screen is usually 60 lines per cm.) and through a reversed photographic positive (Fig. 103; for simplicity we have chosen a grey scale of continuous tone).

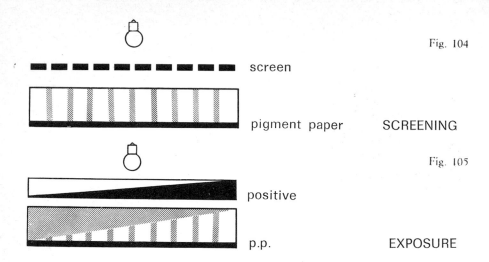

Fig. 104

screen

pigment paper   SCREENING

Fig. 105

positive

p.p.   EXPOSURE

After these two exposures the bichromated gelatin of the pigment paper becames hardened in the areas which have received light (Fig. 104 and 105).

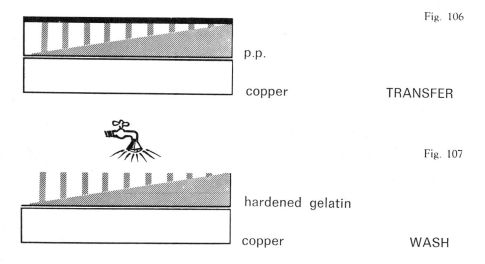

Fig. 106

p.p.

copper   TRANSFER

Fig. 107

hardened gelatin

copper   WASH

The pigment paper is applied (transferred) to the sheet of copper with the gelatin in contact with the copper (Fig. 106 and then, after soaking in hot water, the paper base and the non-hardened gelatin are completely removed (Fig. 107).

The hardened gelatin which remains adhering to the cylinder has a thickness from 2 to 10 microns, the thickness being inversely proportional to the density of the positive.

96

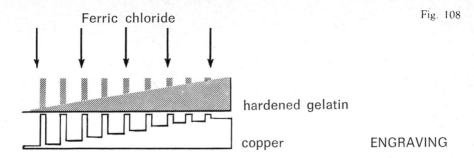

Ferric chloride

Fig. 108

hardened gelatin

copper   ENGRAVING

Fig. 109

copper   ENGRAVED PLATE

By means of solutions of ferric chloride, which penetrate more or less rapidly through the gelatin according to the thickness of this, the copper is dissolved (engraved) more or less deeply (Fig. 108).

After the removal of the hardened gelatin, the small hollows, whose depth is directly proportional to the original positive film, are ready to receive the ink and transfer it by absorption to the printing paper (Fig. 109).

Other systems of engraving the cylinder

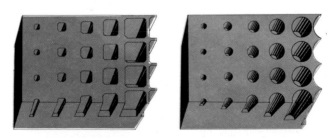

Fig. 110

Fig. 111

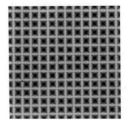

Fig. 112

Fig. 113

97

The engraving procedure we have described is the *traditional* one. It is based, as we already know, on minute cells all of the same area but of variable depth (see Fig. 99).

Today there are also other, modern procedures for engraving the cylinder; the *autotype process* (Fig. 110) with variable area cells but having the same depth; *the Dultgen semi-autotype process* (Fig. 111) with cells having both variable area and depth; *the Respi-Acigraf autotype process* (Fig. 112) with cells varying in area, depth and reduction.

These processes start with a screened positive.

The screening is obtained with special screens (Fig. 113) used as contact screens.

B) **Photographic preparation**

In gravure printing both line and continuous tone originals are subject to screening.

For the printing of line, a line positive of the maximum density is required. The letters of the text must be similarly positive. The base must not exceed a density of 0.15.

For the printing of continuous tone, continuous tone positives are required. Normally the following minimum and maximum densities are called for:

— for Cyan, Magenta and Yellow

minimum	= 0.30
maximum	= 1.70
Scale of D	= 1.40

— for the black

minimum	= 0.15
maximum	= 1.20
Scale of D	= 1.05

It is advisable to use:

1) *for the negative,* a "tone" emulsion of rather soft gradation. This will assure a faithful rendering of the intermediate tones and a good exposure latitude.

2) *For the positive,* whether made by contact or enlargement, a normal "tone" emulsion or, if corrections to the gamma are necessary, one of high contrast.

The Densities may be retouched manually both in the negative and positive, using ferricyanide for reducing them or a black aniline dye for increasing them.

2) OFFSET (Lithography)

A) Method of printing

The printing surface (matrix) is a metal sheet (of zinc, aluminium or bi-metallic) which is perfectly flat and coated with a sensitive emulsion.

After exposure through an original positive, the emulsion is removed from the unprinted areas (Fig. 114) leaving the metal surface uncovered.

Offset printing is based on the phenomenon of repulsion between water and oily substances (inks). The exposed metallic areas of the matrix attract water and repel the ink, while the areas still coated with emulsion attract ink and repel water.

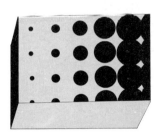

Fig. 114

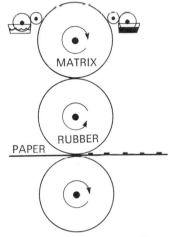

Fig. 115

In bi-metallic plates the two metallic layers are superimposed, the upper one attracting water and the underneath layer attracting oil. After exposure through the original, the upper layer is treated with acid which produces a very light engraving.

In the printing cycle, the image of the matrix (the rightway round) is printed (reversed) on to a cylinder covered with a layer of rubber which in its turn, prints (the rightway round) on the paper (Fig. 115).

B) Photographic preparation

In the lithographic method of printing, the printing areas deposit the ink on the paper always at maximum density. Thus there is no difficulty for line or solid matter, while for continuous tone it is necessary, as we shall see, to make use of screening.

99

1) *For the printing of line,* a line positive of the maximum density is needed made on a lith film.

Letters of text must be equally positive. The density of the base must not exceed 0.20.

2) *For printing continuous tone* we have to resort to the optical device of "screening", that is, to the conversion of the various densities of the image (Fig. 116) into small dots all of the same maximum density but varying in area: the minimum density of the subject (white) is reproduced with very small dots, the maximum with large dots (Fig. 117).

Fig. 116   Fig. 117

It is obvious that, in order for the various densities of a continuous tone original to be reproduced in all their gradations, the size (or area) of the dots must increase progressively from a minimum to a maximum that ranges from 0% to 100%. Practice teaches us however, that the best results in printing are obtained by retaining a very small dot (5%) in the lightest areas and very small "spaces" (95%) in the darkest areas.

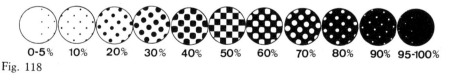

0-5% 10% 20% 30% 40% 50% 60% 70% 80% 90% 95-100%

Fig. 118

This range of dots with area increasing in arithmetic progression is known as the "screen scale", (Fig. 118).

The screen scale and its integral density

The different zones of the screen scale have an overall density which results from the mean between the areas of the dots (black) and the background (white) and is called the "Integral Density". The integral density of the screen image will therefore correspond to the density of a continuous tone and can be easily read with a densitometer.

On page 67 we have shown the ratios between density and dot size (area), which we can now use to plot the following curve (Fig. 119).

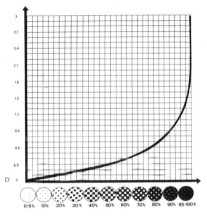

Fig. 119

From it we may observe that an arithmetic increase of the percentage of the dot area or size corresponds to a geometric increase of the integral density.

From this it can be seen that the first half of the screen scale (from 0% to 50%) covers the densities from 0.0 to 0.30, while the second half (from 50% to 99.9%) covers all the other densities from 0.30 to 3.

Screen scale and grey scale

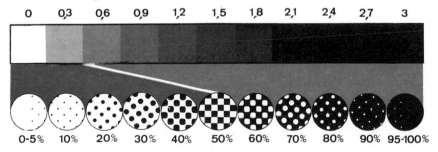

Fig. 120

The same phenomenon can be observed in comparing a screen scale with a grey scale. Placing a normal grey scale with densities from 0 to 3.0 in increments of D 0.3 at the side of a screen scale from 0% to 100% (Fig. 120), we see that the densities from 0 to 0.3 of the grey scale become "expanded" in the first half of the screen scale, with good rendering of detail, while all the other densities become compressed into the second half of the screen scale, with consequent loss in the rendering of detail.

101

Screen scale and the original to be reproduced

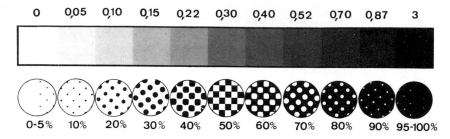

Fig. 121

Fortunately, the continuous tone originals to be reproduced have (unlike the grey scale) their densities distributed in a way very similar to the screen scale, that is, with the densities from 0 to 0.3 already expanded in the first half of the scale, and with the densities from 0.3 to 3 already compressed in the second half of the scale (Fig. 120).

This similarity permits the screening of continuous tone originals readily maintaining the balance of the densities.

The contact screen

The breaking up of the continuous tone original into a screen image for offset printing is generally achieved by means of an "unsharp screen on film" for use by contact.

Fig. 122

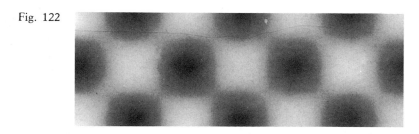

Fig. 123

It is obtained by photographic means from a glass screen (see page 110) and has the dot graduated in density as can be seen in Fig. 122 where it has been enlarged 35 times. In Fig. 123 the "profile of the densities" of the screen dots is shown diagrammatically.

The unsharp screen is used in contact with the lith film, with the emulsion surfaces facing each other. Fig. 123 explains how the screening is accomplished:

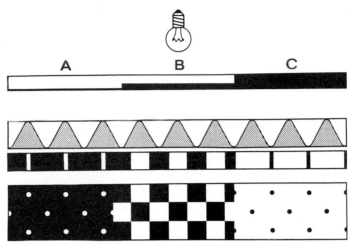

Fig. 124

Where the continous tone original is less dense (A) the light (more intense) succeeds in passing through the maximum density of the dot almost completely and thus blackening the litho film almost completely. Where there are tones of medium density (B) the light succeeds in passing through only the medium densities of the dot, screening the litho film by 50%. Finally, where the density of the original is great (C), the light passes only through the minimum densities of the dot with a screening of barely 10%.

All screens are adjusted for a determined contrast rendering, known as the "basic contrast of the screen", generally of 1.30. That is, with a simple main exposure (original, screen and white light, see Fig. 124), they will give a dot rendering of 5% in the minimum and 95% in the maximum, when the original has a Scale of Densities of 1.30 (Footnote 1).

Whenever the continuous tone original has a Scale of Densities greater or less than 1.30, it will be necessary to decrease or increase the contrast rendering of the screen. This can be done in three ways:

Footnote (1) The basic contrast of the screen is therefore matched *in relation* to the Scale of Densities of the original and is indicated from this. This explains why a screen of *low* basic contrast is indicated by a high number (e.g. 1.40) and is *less contrasting* than a screen of *high* basic contrast, indicated by a lower number (e.g. 1.10).

1 - By means of supplementary exposures

a) A supplementary exposure through the original only (without screen) *increases* the contrast. Such an exposure (called "highlights") should not exceed 15% of the main exposure.

b) A supplementary exposure through the screen only (without original) *decreases* the contrast. Such an exposure (called "flash") is necessary for strengthening the small dots of the screen image.

2 - By controlling the agitation of the development

Vigorous agitation increases the contrast; slow agitation reduces it.

3 - By using a magenta coloured screen

This screen is identical to the grey screen, but has a magenta colour (Fig. 125). Used with yellow or magenta filters, the contrast factor of the screen can be reduced or increased and therefore the contrast of the screen image.

Fig. 125

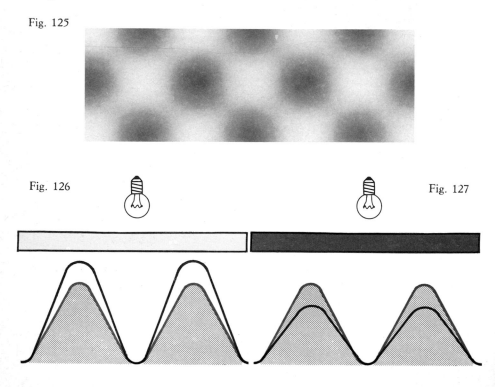

Fig. 126

Fig. 127

The yellow filter in combination with the magenta screen gives a steeper rendering of the density profiles of the dots (Fig. 126), while *the magenta filter* combined with a screen of the same colour, flattens the profile of the dots (Fig. 127).

In this way, if a *supplementary exposure in yellow light* is made after the main white light exposure, the latent dots will be further illuminated *in their central part* with the enlargement of dots of low percentage (5% to 50%) without any effect on the high percentage dots (50% to 95%). This will obviously produce a lowering of the contrast in the screen image up to a basic contrast of 1.70.

On the other hand, il *a supplementary exposure is given in magenta light* after the main white light exposure, the latent dots will be further illuminated *in their outer zones* with enlargement of the high percentage dots (50% to 95%), but without any effect on the low percentage dots (5% to 50%) which are protected by the density of the original. An increase in basic contrast up to 0.90 can be obtained in this way.

Polichrom "yellow" and "blue" screens

These make use of the same technique as the magenta screen

Fig. 128   yellow screen   blue screen

The yellow screen is for use with tungsten light sources which are rich in yellow-red radiation; *the blue screen* with xenon lamps which are rich in violet-blue radiation.

Exposed with white light, *the yellow screen,* has a very high basic contrast (0.90); used in blue light (the complementary of yellow) it has a very low basic contrast (2.10). By giving varying amounts of the two lights, perfect screen images (dots from 5% to 95%) can be obtained from originals with Density scales of from 0.90 to 2.10.

The blue screen gives the same results if used in white light (basic contrast = 0.90) or in yellow light (basic contrast = 2.10).

The advantages of these screens are:

1) The very short exposure required (about 1/3 of that for a magenta screen).
2) The use of only one colour filter: either yellow or blue (instead of two yellow and magenta — required by a magenta screen).
3) The wide range of correction of the Density scale of originals.
4) The extreme facility with which, (by means of a preliminary test), the exposures in white and coloured light can be pre-determined.

Screens for positives and negatives. The shape of the dot. Technical data. Angle of dots.

For the preparation of screened positives or negatives, a study has been made of suitable screens which favour the progression of tones on the basis of the printed results (Fig. 129).

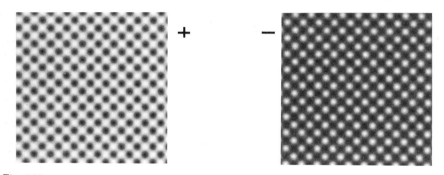

Fig. 129

Screens are classified according to the following data:

— format: e.g. 18 x 24 cm., 60 x 80 cm., 90 x 90 cm. etc.
— angle of the dot pattern: 45° - 75° - 90° - 105° (see Fig. 131)
— fineness of dot pattern: 20, 40, 54, 60, 80 dots (or lines) per cm. (65, 85, 133, 175, etc. per inch)
— basic contrast: 1.20, 1.30, 1.40, etc.
— colour: grey (neutral), magenta, yellow, blue.
— shape of dot: "square", "elliptic", "respi", and "respi-elliptic", as shown in Fig. 130 A B C D.

The normal "shape" of the dot is that the "square" (Fig. 130 A).

There are, however, screens with dots of "elliptical" shape in commercial use (Fig. 130 B): these give a softer reproduction, especially in the medium tones. Screens obtained with the superimposition of two screens of equal fineness as shown in Fig. 30 C ("respi" screens); up to 20% of the dot strength only the large dot pattern is effective, and beyond 20%, the second and finer pattern comes into play giving an extraordinary rendering of detail of the original. A variation of this screen having dots of "elliptical" shape (Fig. 30 D) gives both softness and rendering of detail.

Fig. 130

In respect of the "angle" of the dot pattern of the screen, we illustrate schematically the optimum angles (Fig. 131) for avoiding the "moiré" effect which can arise when parallel lines of dots are superimposed.

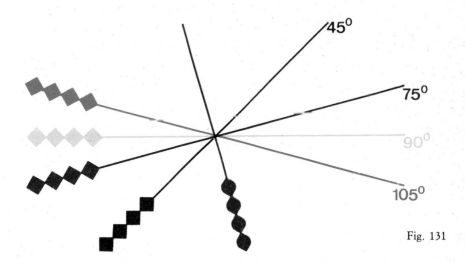

Fig. 131

The "chain" of dots given by the Black printing must have an angle of 45° in relation to the vertical; that of the Magenta of 75°; that of the yellow of 90°; and that of the Cyan of 105°. When using screens having elliptical dots (see Fig. 130 B and D) the angle of the Magenta must be rotated through 90° in respect to the normal of 75°.

Fig. 133 shows a greatly enlarged section of the picture in Fig. 132, where, by half-closing the eyes, it is possible to see the characteristic "rosette" pattern of properly oriented (angled) screens.

Fig. 132

Fig. 133

3) LETTERPRESS

A) Method of printing

The printing surface (matrix) is a plate of metal (usually of zinc or copper) with some areas in relief in respect to the plane of the plate (Fig. 134).

The areas in relief all have the same height, but not the same area, for which reason continuous tones must be reproduced, as in lithography, by means of a half-tone screen image.

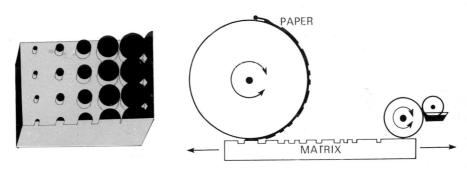

Fig. 134   Fig. 135

In the printing cycle the ink is deposited on the parts of the matrix in relief by inking rollers. The matrix carrying the ink is then moved under a pressure roller in contact with the paper to which it tranfers the ink (Fig. 135).

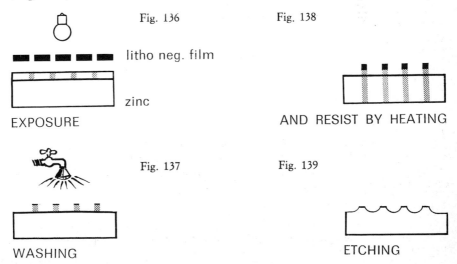

Fig. 136   Fig. 138

litho neg. film

zinc

EXPOSURE   AND RESIST BY HEATING

Fig. 137   Fig. 139

WASHING   ETCHING

The relief areas of the plate are obtained in the following manner: the zinc plate is coated with a layer of polyvinyl alcohol sensitised with ammonium bichromate (footnote 1) and is then exposed to light through a reversed (right-reading) negative, either line or screened (Fig. 136). In the parts receiving light the polyvinyl alcohol becomes insoluble and the unexposed areas are removed by washing with water (Fig. 137). The plate is then hardened in a chromic acid bath before being heated in an oven to 200-230° C. The polyvinyl alcohol image changes to form an extremely hard acid-resisting layer.

Etching is now carried out in a solution of nitric acid which dissolves the unprotected areas of the zinc in the non-printing parts leaving a relief image (Fig. 139).

The etching can be carried out in a dish or with special automatic machines (Dow-Etch method); in this case, if we are dealing with a screened image, the percentages of the dots will have to be modified, as we shall explain later.

B) Photographic preparation

As already mentioned, in letterpress printing continuous tone subjects must first be screened, as in lithography.

Although the screening can easily be obtained by means of a contact screen, it is more common to use optical screening by means of a glass screen employed in a process camera.

According to the method of etching, *the screened negative* should normally have the following dot percentages:

Dish etching:

Minimum dot percentage (dark areas of original): 5%
Maximum dot percentage (light areas of original): 60%

Machine etching:

Minimum dot percentage (dark areas of original): 15%
Maximum dot percentage (light areas of original): 80%

Optical screening

The screened negative for making letterpress matrices is, as we have already said, generally obtained by means of optical screening.

In this case the breaking up of the continuous tone into so many small dots of different area is accomplished as follows:

Footnote (1) Zinc plates are available commercially which are pre-sensitized and do not require the stages of sensitising, and heating.

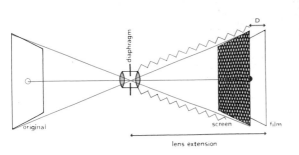

Fig. 140

Fig. 141

A special glass screen is placed in front of the film inside the process camera (Fig. 140). The screen is formed by two sheets of optical glass on whose inner faces have been etched very fine parallel grooves which are filled with black pigment or magenta dye. The two sheets are cemented with the grooves in contact and perfectly at right together angles to each other. Normally the small trasparent squares which result are of the same width as the black lines: in this case the screen is said to have a ratio of 1 : 1. The "fineness" of the lines, that is the number of the lines per cm. can be 20, 30, 40, 60, 70 and even greater. The appearance of the screen, greatly enlarged, is shown in Fig. 141.

Let us now return to Fig. 140. The light reflected by the original passes through the lens of the process camera as in a normal photographic exposure.

However, before reaching the film, the light passes through the glass screen, and it is here that it becomes modified, dividing itself into so many small luminous rays.

Every small transparent square of the screen behaves in fact like a tiny lens which collects an equally small part of the original, transforming it into a luminous ray which will affect a small part of the film. It is clear that the dot which will be formed on the film will be larger or smaller in proportion to the intensity of the small ray.

The distance of the screen from the sensitive film is given by the following equation:

$$D = t \frac{s}{d}$$

where D is the distance between the black line screen and film in millimetres;

t is the lens extension of the camera in mm.;

s is the side of the square dot of the screen in mm.;

d is the diameter of the lens diaphragm in mm.

To obtain dots of the required size in the maximum and minimum "densities" of the screened negative, various techniques are employed which we shall only briefly touch on:

1) Increasing the distance of the screen from the film increases the contrast and vice versa.

2) Increasing the aperture of the lens diaphragm increases the contrast and vice versa.

3) The smaller dots of the screened negative can be reinforced by giving a brief supplementary exposure with a sheet of white paper covering the original. This is in addition to the main exposure with the original and is referred to as the "flash" or "pre-exposure".

4) By making a second very short supplementary exposure of the original but with the lens diaphragm widely opened (known as exposure for "highlights") the saturation is increased in the darker parts of the screened negative (i.e. the "highlights" of the original).

4) ELECTROSTATIC PRINTING

A) Method of printing

A very modern method of printing, which is seen by technicians as having considerable development in the future, is that based on the electrostatic deposition of the inks.

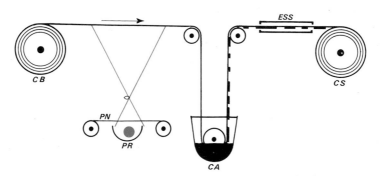

Fig. 142

The white paper CB (Fig. 142) is previously coated with a thin layer of zinc oxide, a substance which does not conduct electricity in the dark but which, exposed to light, becomes a good conductor.

An optical projector PR projects an image of the negative PN (matrix) on to the paper which becomes electrically charged in the zones struck with light with the formation of an electric latent image.

The passage of the paper through a trough CA containing ink charged electrically with a charge opposite to that of the paper, will determine (on the principle explained in Footnote 1 on page 15) the deposition of ink on the parts of the paper previously illuminated.

The ink is then fixed on the printed paper CS by passing through a drier ESS.

By successively repeating this operation on the same sheet it is possible to obtain printing in full colour.

B) **Photographic preparation**

The matrices are transparent negative films and their preparation is identical to that used in lithography.

COLORIMETRY

PART II

COLORIMETRY

The reproduction by printing of a coloured original, presupposes a knowledge of the techniques which apply to the reproduction of a black and white original, which we have dealt with in Part I. In fact, it is done by successively printing in register, one above the other, the different colours as if they were — taken singly — monochrome originals.

To this knowledge it is necessary to add something on the theory of colour and those practices concerned with the making of colour separations and the methods of colour correction.

This second part is therefore divided into three chapters:

1 Colour

2 The separation of colours

3 The correction of colours

CHAPTER I

COLOUR

In this first chapter we shall briefly expound the theories of colour where they are indispensible to an understanding of the various techniques of separation and masking to be described in chapters 2 and 3.

The treatment of this subject will be subdivided into four sections: 1) Colour - light; 2) Colour - pigment; 3) Characteristics and specification of colour; 4) Light sources and the photographic emulsion.

1) COLOUR - LIGHT

a) Light

Light is one of the many forms of "radiant energy", that is, energy which is propagated by means of electromagnetic waves.

Wave propagation is characterised by two factors: wavelength and frequency.

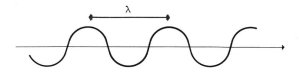

Fig. 143

The wavelength (λ, read "lambda", which is the Greek L) is the distance between two adjacent wave crests (Fig. 143).

The frequency (n) is the number of waves which pass a point in one second:

$$n = \frac{n_o \text{ of waves}}{\text{sec.}}$$

From this we deduce that the *velocity* (v) of the wave propagation is given by the product of the wavelength and its frequency:

$$V = \lambda \times n$$

Experience shows that the velocity (v) of radiant energy varies according to the medium through which it passes. For example, the velocity of light which is 186,000 miles a second (300,000 km/sec.) in vacuum, is reduced by one third in glass becoming about 124,000 m.p.s. (200,000 km/sec.).

This means, that in glass, either the wavelength of the light or its frequency or both will diminish.

In fact, it is the wavelength which diminishes, while the frequency remains unchanged (Fig. 144).

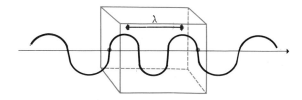

Fig. 144

118

However, since it is easier to measure the wavelength than the frequency, it is more convenient to identify a given luminous radiation in terms of its *wavelength measured in air*.

Fig. 145 is a simplified diagram showing the principal forms of radiant energy in what is known as the *electromagnetic spectrum*.

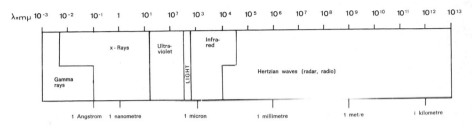

Fig. 145

Wavelength (λ) is measured in *millimicrons* (Footnote 1). The micron (μ) is the thousandth part of the millimetre ($1\mu = \dfrac{mm.}{1000}$) and the millimicron ($m\mu$) is the thousandth part of a micron ($1m\mu = \dfrac{\mu}{1000}$).

Sometimes the very short wavelengths are measured in the Angstrom ($\mathring{A}$) unit which is the tenth part of a millicron ($1\mathring{A} = \dfrac{m\mu}{10}$).

The known forms of radiant energy extend from the gamma rays (very short wavelengths where $\lambda = 10^{-3}$ mμ), to the hertzian (radio) waves reaching wavelengths of many kilometres ($\lambda = 10^{13}$ mμ).

The form of radiant energy *visible* to the eye (which we call *light*) extends from about 400 to 700 mμ. It is preceded by ultra-violet radiation and followed by infra-red in the electromagnetic spectrum, but since both are invisible, they do not merit the term "light".

Light can thus be defined *as that type of radiant energy* (objective aspect or physical) *visible to the eye* (subjective aspect or psychological). Light is therefore a psycho-physical phenomenon.

Footnote (1) The millimicron is also known as the nanometre with the symbol nm.
As noted, numbers multiplied or submultiplied by 10 can be written in the form of powers to the base of 10 and with exponents of so many units as there are zeros in the number. The exponent is positive for multiples and negative for submultiples of 10.
Thus $10 = 10^1$; $100 = 10^2$; $1000 = 10^3$ etc.
while $\dfrac{1}{10} = 10^{-1}$; $\dfrac{1}{100} = 10^{-2}$; $\dfrac{1}{1000} = 10^{-3}$ etc.

b) The light spectrum

As stated above, the visible electromagnetic spectrum, that is, the luminous spectrum, extends from 400 to 700 mμ.

When *all* the wavelengths from 400 to 700 mμ strike the human eye at *the same time* and in about the *same quantity*, that is, with the same intensity, we receive the sensation of *white light*.

On the other hand, when the eye receives only *a portion* of the visible spectrum (for example, only the wavelengths from 400 to 500 mμ), we have the sensation of *coloured light* (in this case blue).

This means that coloured light is a part of white light.

This can be demonstrated experimentally by passing a beam of white light through an optical prism (Fig. 146).

Fig. 146

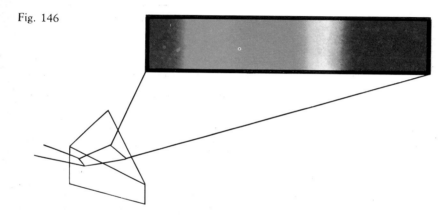

By an optical process which need not concern us here, white light (400 to 700 mμ) passing through a prism is dispersed into its various wavelengts, making the colours of which it is composed visible. This gives us the "visible spectrum" of light.

Cyan Yellow

Fig. 147

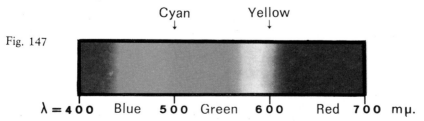

λ = 400 Blue 500 Green 600 Red 700 mμ.

In the light spectrum the three principal coloured lights are: *blue-violet* (about 400 to 500 mμ), *green* (about 500 to 600 mμ) and *red* (about 600 to 700 mμ), which are the so-called *primary colours of light* (Fig. 147).

c) **The additive synthesis (of coloured light)**

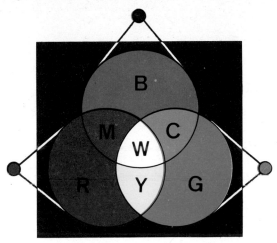

Fig. 148

By projecting lights of the three primary colours so that the three overlap as shown in Fig. 148, the area where all three overlap (at the centre) will give a recomposition of white light.

In those zones where only two of the colours overlap, new colours, nearer white but without arriving at it, will be formed: the blue and green lights give *cyan* (C) light; the green and red given *yellow* (Y) light; the red and blue give *magenta* (M) light. These are sometimes known as the *secondary coloured lights*. This "addition" of coloured light which approaches (zones M Y and C) or in fact arrives at white light (centre) is called the "*additive synthesis*" of coloured lights (Footnote 1).

d) **The complementary colours of light**

Referring again to Fig. 148 we shall notice that white light (at the centre) can be reconstructed, not only by the superimposition of the three primaries, but also by superimposing one of the primary colours (for, example blue) with one of the secondary colours (for example, yellow) resulting from the sum of the other two primaries (green and red). This secondary coloured light (yellow) is referred to as the "complementary" colour to the primary colour (blue).

Footnote (1) Cyan light, being the addition of blue and green light is also called blue-green. Similarly, magenta can be referred to as blue-red and yellow as green-red.

The colours of cyan and yellow can be seen in the light spectrum — see Fig. 147. Magenta light is not seen in the spectrum, but is reproducible artificially as the « complementary » of green light.

The complementary colours of light are therefore none other than those which, being mixed with another (whether primary or secondary) give white light.

We can therefore view the complementary colours as follows:

PRIMARY COLOURED LIGHT	BLUE ●	GREEN ●	RED ●
ITS COMPLEMENTARY	YELLOW ○	MAGENTA ●	CYAN ●

Fig. 149

2) COLOUR - PIGMENTS

a) Pigments

The objects which surround us appear coloured to our sight because they are constituted of or covered with chemical substances called *pigments,* which have the property of absorbing some or all of the wavelengths of light and reflecting back the rest to our eye.

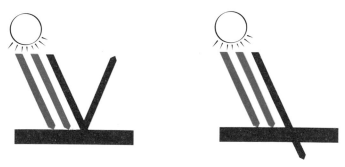

Fig. 150

Fig. 151

For example, an object appears red to us (Fig. 150) because the white light (blue + green + red) is partially absorbed (blue + green) and partially reflected (red) by the "pigment" of the object.

The same phenomenon applies if the pigment is incorporated in a transparent substance such as a sheet of glass.

The glass appears red to us (Fig. 151) because the white light (blue + green + red) is partially absorbed (blue + green) and partially transmitted (red) by the pigment or dye in the glass.

There are thus types of objects that we see as coloured because of the "pigment" which they incorporate or are covered with: those which are *opaque* (to which belong some of the inks used in printing) and those which are *transparent* (to which belong the colour "filters").

The first *reflect* and the second *transmit* only the wavelengths that we see, *absorbing* the others. This phenomenon is referred to as "selective absorption".

Let us note immediately that while speaking of "coloured light" we have designated as "primary" Blue, Geen and Red light because they stand at the base of the additive synthesis of light, but speaking of "*coloured pigments*", we shall call the colours Magenta, Cyan and Yellow the "primary colours" or "basic colours" because it is these colours which *form the basis of the subtractive synthesis of colours.*

b) **Subtractive synthesis (of the colour-pigments)**

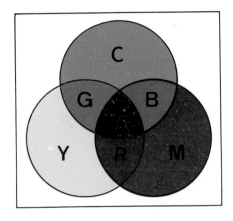

Fig. 152

By overlapping the transparent filters of Cyan, Magenta and Yellow (Fig. 152) over a viewing screen of diffuse white light, some of the wave lengths of white light will be absorbed by the dyes of the filters and only the transmitted wavelengths will reach our eye, as indicated in the table below:

Fig. 153

			LIGHT ABSORBED	LIGHT TRANSMITTED
Zone C	CYAN	(= green + blue)	RED	GREEN and BLUE
Zone M	MAGENTA	(= red + blue)	GREEN	RED and BLUE
Zone Y	YELLOW	(= red + green)	BLUE	RED and GREEN

123

Fig. 154

In zone C (Cyan), the transmitted light will be that of Blue and Green (in fact in the additive synthesis, Blue + Green light give Cyan, as already noted on page 121) and the light absorbed will be Red (Fig. 154).

Fig. 155

In zone M (Magenta), the transmitted light will be Red and Blue, and that absorbed will be Green (Fig. 155).

Fig. 156

In zone Y (Yellow), the transmitted light will be Red and Green and the absorbed light Blue (Fig. 156).

In the zones in which the filters overlap, the transmitted light will be that light which both the filters transmit, that is:

Fig. 157

In zone C + M the transmitted light will be Blue because Blue light is trasmitted by both the Cyan and Magenta filters, while Red and Green are absorbed, the first by the Cyan and the second by the Magenta filter (Fig. 157).

Fig. 158

In zone M + Y the transmitted light will be, on the same basis, that of Red (Fig. 158).

Fig. 159

In zone Y + C the transmitted light will be that of Green (Fig. 159).

Fig. 160

The overlapping of all three filters (Cyan, Magenta and Yellow) will not transmit any wavelength of white light and for this reason the centre zone will appear black (Fig. 160).

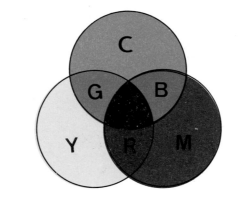

Fig. 161

Fig. 162

The same thing will happen by printing, one partially or wholly super-imposed on the other (Fig. 161) or by mixing together (Fig. 162), three inks having the colours of Cyan, Magenta and Yellow (the three basic colour-pigments).

Fig. 163   Fig. 164   Fig. 165

Basic or primary colour pigments (inks): Cyan (Fig. 163); Magenta (Fig. 164); Yellow (Fig. 165).

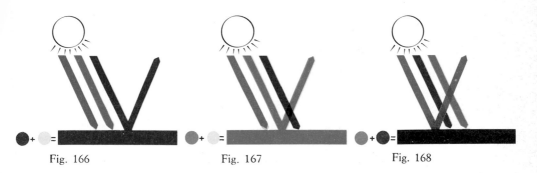

Fig. 166   Fig. 167   Fig. 168

Secondary colour pigments (inks): Red (Fig. 166); Green (Fig. 167); Blue (Fig. 168).

Fig. 169

Colour pigment (ink) black (Fig. 169).

This subtraction from white light effected by colour pigments extending as far as giving black, is called "subtractive synthesis".

c) The complementary colour pigments

We have seen that coloured lights overlapping in additive synthesis lead to white and that colour pigments overlapping in subtractive synthesis lead to black.

We can thus say that two colour pigments are "complementary" to each other when, overlapping (in subtractive synthesis) they give black, as shown below:

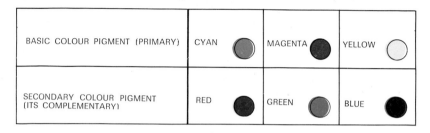

Fig. 170

3) CHARACTERISTICS AND SPECIFICATIONS OF COLOUR

a) Psychological characteristics (subjective) of colour

When we "see" a colour, the sensation we have of it is characterised by three values: hue, saturation and luminosity.

The hue of a colour is its "qualitive characteristic": thus we have the hues of red, green, yellow, etc.

The saturation of colour is its "quantitative characteristic": a hue is the more saturated *the less white it contains.* For example, red is more saturated than pink, because it contains less white than pink.

The luminosity (or brightness) of a colour is its "reflection characteristic": a hue is the more luminous the *less black (or grey) it contains.* For example, light red is more luminous (brighter) than dark red because it contains less black and hence reflects (or transmits in the case of a transparent medium) more light.

Hue, saturation and luminosity are independent of each other. We can see two different hues (for example, red and green) with the same saturation and/or luminosity, or two identical hues (for example, reds) with different saturations and/or luminosities, (Footnote 1).

Footnote (1) Another subjective valuation of colours classes them as « warm » or « cold ». Yellow and red, the colour of fire, are generally classed as « warm », while the colour of water (green, cyan and blue) are considered « cold ». A green could be « warm » if it is yellowish, « cold » if bluish.

In printing, the characteristics of colours are rendered as follows:

the hue depends of the wavelength of the colour of the ink;

the saturation depends on "the dot area" (or in gravure on the depth of the cell);

the luminosity depends on the "purity" of the colour of the ink and on the "whiteness" of the paper.

b) Physical characteristics (objective) of colour

The scientific measurement (objective) of colour is rendered possible by special instruments known as spectrophotometers in which the colour to be measured is compared with the electromagnetic spectrum of a standard source of white light.

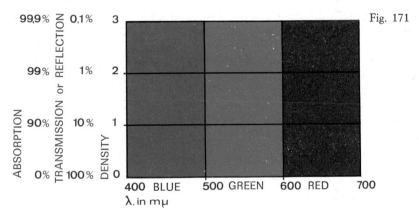

Fig. 171

Fig. 171 represents a simplified absorption spectrum of a standard white surface: the horizontal axis is marked with the wavelengths of Blue ($\lambda = 400\text{-}500$), Geen ($\lambda = 500\text{-}600$) and Red ($\lambda = 600\text{-}700$), while the vertical axis is scaled with densities from 0 to 3 or (amounting to the same thing) the percentages of transmission (or reflection) from 100% to 0.1%, to which the percentage absorptions from 0% to 99.9% correspond. From this it results that a perfect white surface has an absorption equal to zero and a reflection of nearly 100% at all the wavelenghs of the visible spectrum, from 400 to 700 mμ).

Now let us assume that we must obtain from a spectrophotometer, the absorption spectrum of a red coloured object: this, as we know, absorbs all the Blue and Green radiation (400-600) and transmits the Red (600-700), thus forming a curve (Fig. 172) called the "spectral absorption curve", which is characteristic of the colour red (in this case a Kodak Wratten No. 25 red filter).

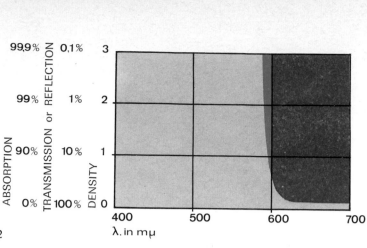

Fig. 172

λ. in mμ

The grey zone of the diagram bounded by the curve is that of the wave-lengths "absorbed", the red that of the wavelengths "transmitted".

The curve tells us that even in the red zone from λ 700 to about λ 620, there is an absorption of 15% (85% Transmission, 0.07 Density); that at λ 600, the absorption is 50% (50% Transmission, Density 0.3) and lastly that the total absorption (0.1% Transmission, Density 3) occurs at λ 590.

In this way any colour can be graphically represented by its spectral absorption curve, and recognised at a glance.

c) Specifications of colour

One of the most important results to which our study of colour should conduct us is to be able to define every colour in objective terms, valid for everyone.

Such a system which has now been universally adopted is that known as the CIE, so named after the Commission Internationale de l'Eclairage.

Fig. 173

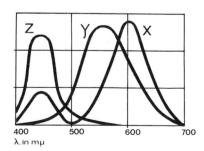

λ. in mμ

130

All colours can be matched by mixtures of three primarics and the CIE system is based on three theoretical primaries, mixtures of which can be found that match all the spectral colours. The quantities of these primaries required to match a particular colour are called "tristimulus values" and have been given the symbols X, Y and Z. Figure 173 shows the values of X, Y and Z for the spectral colours (Footnote 1).

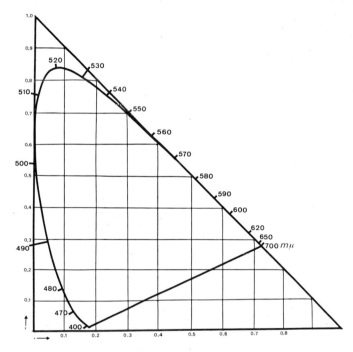

Fig. 174

These three colours (X, Y, Z) contribute in different proportions to all the other colours but in such a way that the sum of their amounts (indicated by the corresponding lower case letter x, y and z and called "trichrometric coordinates") is always equal to 1:

$$x + y + z = 1$$

Hence this equation can also be written so:

$$z = 1 — (x + y)$$

In this way it is possible to consider only the pair x and y and hence to trace the curve joining all the wavelengths of the spectrum colours, from 400 to 700 mμ (Fig. 174).

This curve has the shape of a horseshoe with the toe above and can be enclosed in a triangle, which is called the "CIE triangle".

Footnote (1) The tristimulus values X, Y, Z, are so chosen that Y not only represents the theoretical colour Green, but also the luminosity.

Fig. 175

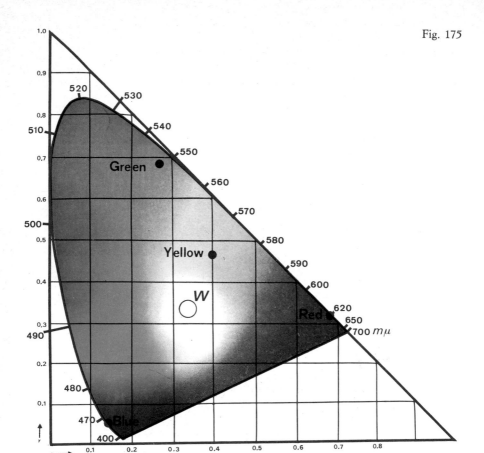

The colours enclosed by it (Fig. 175) are all formed from the three theoretical colours X, Y, Z so combined that every point of the area of the triangle observes the equation x + y + z = 1.

For example, a Red x = 0.68; y = 0.32; will have z = 0 because
    0.68 + 0.32 + 0 = 1.
A Green x = 0.27; y = 0.68; will have z = 0.05 because
    0.27 + 0.68 + 0.05 = 1.
A Blue x = 0.14; y = 0.05; will have z = 0.81 because
    0.14 + 0.05 + 0.81 = 1.
A Yellow x = 0.40; y = 0.47; will have z = 0.13 because
    0.40 + 0.47 + 0.13 = 1.

The point W, in which x, y, and z are all equal to 0.333 (the sum of which is 1) is called "white of equal energy".

Every colour will in this way be specified *from its position in the CIE triangle,* by means of the two colorimetric coordinates x and y.

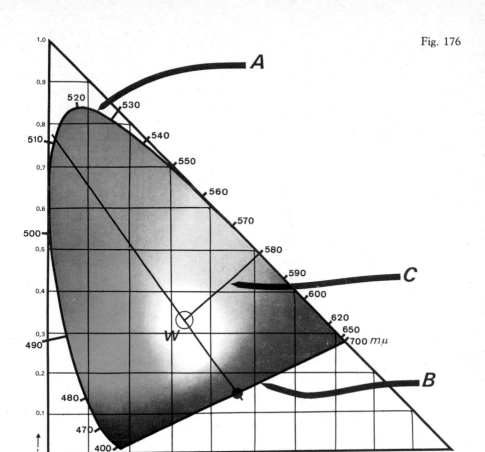

Let us now examine the CIE triangle (Fig. 176), from which we wish to obtain the principal specifications of colour.

1. - The various points on the horse-shoe curve (A) represent the various hues, that is, the spectrum colours corresponding to the wavelengths marked at the side.

This specification is known as the *"dominant wavelength"*.

The points of the straight line (B) which joins the extremities of the curve (called the line of purple) represent the colours not existing in the spectrum, but which are artificially reproducible. They are indicated by the wavelength of their complementaries preceded by a minus sign (—) or by a C. Thus the Magenta, which is complementary to Green 511 will be shown by the number — 511 or C 511.

2. - The various points of the straight line (C) which joins any point of the horse-shoe curve with the point W represent colours of equal hue but of decreasing *saturation* as they get closer to W.

The maximum saturation is on the curve. By dividing the distance from the point W to the point of the colour in question by the total distance from W to the curve, we obtain a measurement of the purity of the colour, which will be at a maximum on the curve and nil at point W.

This specification is called *"excitation purity"* and is expressed as a percentage.

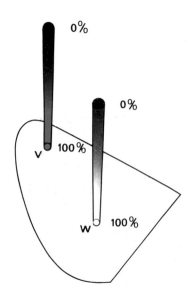

Fig. 177

3. - The length of the imaginary perpendicular erected from the point of the surface that represents the colour in question, gives us, as a percentage, *the luminosity value* of the hues of the various points.

In the CIE triangle, which is a flat (two-dimensional) figure, only the hue and saturation are represented, while the luminosity cannot be graphically shown. It can, however, be visualised by referring to Fig. 177 (tri-dimensional). The two columns of green and white which are erected over the points of two hues (green x = 0.15; y = 0.75 and white W) have a luminosity of 100% at the base and 0% (black) at the summit (Footnote 1).

We can thus specify also the luminosity of a hue of given saturation by means of a number which expresses the percentage of it.

This specification is called the *percentage reflection or transmittance*.

Footnote (1) We note that the column that rests over W is a true and proper scale of greys.

134

Let us now give a concrete example of how colour should be specified by the CIE system.

This also represents a summary of what we have so far said.

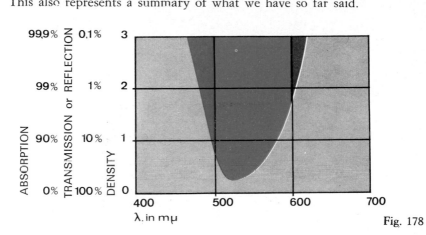

Fig. 178

A colour which has been submitted to a spectrophotometric analysis, has given the absorption curve in Fig. 178. This colour is the Green of the Kodak Wratten No. 58 filter.

We can give the specification as follows:

Dominant wavelength (hue) = 540.3 mµ
Purity excitation (saturation) = 86.2%
Luminous transmittance (luminosity) = 23.7%
CIE position: x = 0.24; y = 0.70

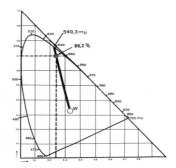

Fig. 179

The CIE position has been depicted graphically (Fig. 179) by first marking the dominant wavelength on the curve (λ = 540.3; then by tracing a straight line from this point of the curve to the point W and fixing to the excitation purity of 86.2% the point of the colour. The two coordinates drawn from the colour point to the axes x and y (x = 0.24; y = 0.70) are the position of the colour in question in the CIE triangle.

4) LIGHT SOURCES AND THE SPECTRAL SENSITIVITY OF THE EMULSION

The knowledge that we have now acquired about colour will enable us to understand quite easily the spectral characteristics of light sources and photographic emulsions and the relationships between them.

a) Light sources

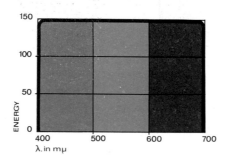

Fig. 180

The sun. The light that it emits is almost white, that is, it contains *nearby equal amounts* of *all* the visible radiations. Its "spectral energy" curve is reproduced in Fig. 180 in simplified form by the black line. In it, the luminous energy is hown in relative values on the vertical axis against the wavelength along the horizontal axis.

Sources of artificial light rarely approach the perfection of the sun. In some the light emitted contains only a *limited number* of the visible radiations, while in others all the radiations are present but *not in equal proportions*.

Let us examine some of the principal light sources:

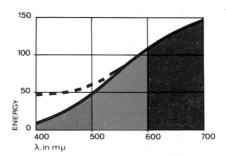

Fig. 181

— *Incandescent lamps* - These are the normal tungsten filament lamps used for room illumination. Their spectral emission (Fig. 181) is of the "continuous type" (that is, without distinct "bands" of radiation, as we shall find in other lamps).

136

It shows us that an incandescent lamps emits light in *all* wavelengths, but *not in equal amounts,* giving more red light than blue. It is for this reason that tungsten lighting appears yellower than sunlight.

This deficiency in blue is partly corrected by operating tungsten lamps at a much higher voltage (shown by the broken line in Fig. 181), typical lamps being the well-known "photofloods" used in the photographic studio. However, such lamps have a very much reduced burning life.

— *Mercury vapour lamps* - These are electric discharge lamps consisting of a glass tube containing mercury vapour (without filament).

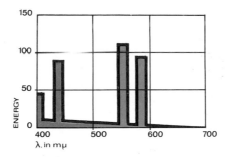

Fig. 182

Their spectral emission takes the form of "bands" (Fig. 182).

The curve tells us that a mercury vapour lamp emits *only some wavelengths,* chiefly in the region of ultra-violet and blue-green radiation.

— *Fluorescent tubes* - These are also discharge lamps. The internal surface of the tube is covered with a substance, which, when struck by the

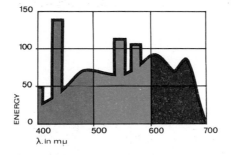

Fig. 183

ultra-violet radiation emitted by the gas, becomes fluorescent emitting visible light.

Their spectral emission (Fig. 183) is a combination of continuous and band emissions.

The curve tells us that fluorescent tubes emit light in *all* the visible wavelengths but *not in equal proportions.*

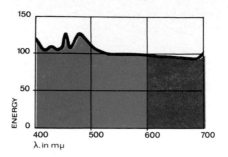

Fig. 184

— *Xenon lamps* - These are electric discharge lamps containing the rare gas xenon at high pressure.

Their spectral emission (Fig. 184) is of the "continuous" type.

The curve tells ut that the lamp emits luminous radiation of *almost equal amount* in the whole visible spectrum.

Hence their light closely approaches that of ideal sunlight.

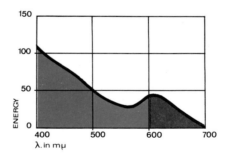

Fig. 185

— *Arc lamps* - These are based on two carbon electrodes between which a discharge of electricity takes the form of a small arc. The light emission comes from the arc and from the glowing colour.

Their spectral emission (Fig. 185) is of the continuous type.

It tells us that the arc lamps emits *all* the luminous radiations but with a *preponderance in the blue region* of the spectrum.

— *Electronic flash* - This has similar spectral emission to the xenon lamp (see Fig. 184), being itself a glass tube filled with xenon gas. However, the electric discharge is very short (in the region of 1/500 sec.) but of very high intensity. With certain emulsions it may be necessary to take into account the Schwarzschild effect (see page 50).

b) The colour temperature of light sources

We already know — at least theoretically — that black absorbs all the radiations that strike it and thus reflects none.

If we now assume that we have a body that effectively absorbs *all* the radiation (the so-called "black body"), by heating it (as in the case of our iron bar) it will first emit light of a dull red, then as the temperature is increased, becomes bright red, yellow, white and finally blue.

The colour of the light is therefore closely related to the temperature: that is, there is a colour-temperature ratio.

Colour temperature is measured in degrees Kelvin (°K), a scale which starts with absolute zero which is equal to — 273° Centigrade.

Colour temperature (in degrees Kelvin) is therefore equal to the temperature in centigrade + 273

$$°K - °C. + 273$$

Footnote; It is now universally agreed to discard the degree sign in specifying colour temperature and, for example, 5,000° K. is now written 5,000 K.

Fig. 186

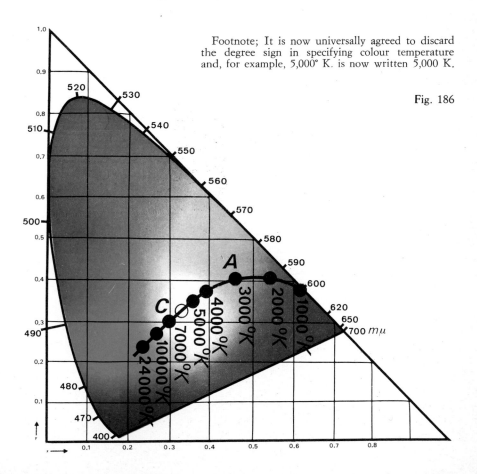

139

Fig. 186 shows the position of various colour temperatures in the CIE triangle.

A. at 1,000 K the light is red (600 mμ).

A. at 2,000 K the light is orange (585 mμ).

A. at 3,000 K the light is orange-yellow (580 mμ). This, at 2,854 K, corresponds to the "Standard Source A" (tungsten light)

A. at 4,000 K the linght is white. At 6,750 K, it corresponds the "Standard Source C" (artificial daylight)

From 7,000 to 10,000 K the light is faintly bluish

From 10,000 to 24,000 K the light is sky blue.

To give some examples, a normal tungsten lamp has a colour temperature of 2,800 K; photoflood lamps (for photography) are 3,200 or 3,400 K; arc lamps in the region of 5,000 K; while xenon lamps and electronic flash are in the region of 6,300 K. "Daylight" type fluorescent tubes are about 6,500 K.

Generally speaking we can say that *the colour temperature increases as the blue component (shorter wavelengths) increase in relation to the red (longer wavelengths)*, as shown in Fig. 187.

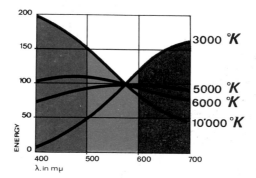

Fig. 187

The curves also tell us that light sources with the most balanced spectral distribution are those between 5000 and 6000 K.

c) Spectral sensitivity of photographic emulsions

It is necessary to distinguish between the effect that a given light source has on the human eye (luminous effect) and that produced on a photographic emulsion (actinic effect).

Photographic emulsions have, in fact, spectral sensitivites which can differ very much from that of the eye. They may be sensitive to certain wavelengths of the solar spectrum to which the eye is insensitive and vice versa.

Let us look at the sensitivity of various types of photographic emulsion in terms of wavelength:

140

1) Non colour sensitive emulsion (or blue-sensitive)

The silver halides (for example, silver bromide, AgBr) are, by nature, only sensitive to large energy radiations, that is to blue-violet radiation (400-500 mμ) and ultra-violet (as far as 250 mμ), but are insensitive to radiations of less energy such as green, yellow and red.

These emulsions with silver bromide in the natural state are called "non-colour sensitive" or "blue-sensitive" or "ordinary" emulsions.

The curve of their sensitivity to the various wavelengths is usually represented by the so-called wedge *spectrogram*, obtained by exposing the emulsion under test to the luminous spectrum given by a prism. The wavy horizontal lines indicate the points of equal illumination against the horizontal scale of wavelengths.

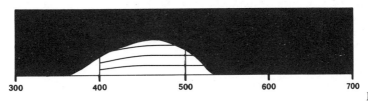

Fig. 188

Fig. 188 shows the spectrogram of a "blue-sensitive" emulsion. Emulsions of this type can be safely handled in a yellowish-green or red light.

2) Orthochromatic emulsions

The crystals of silver bromide can be made sensitive to radiation of less energy by treatment with dyes (colour-sensitising) which, by absorbing the radiation, transmit it to the crystals, thereby forming the silver of the latent image.

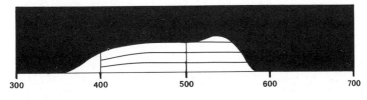

Fig. 189

In this way it is possible to obtain emulsions which are also sensitive to green and yellow-green light (500-600 mμ), called "orthochromatic" emulsions. The spectrogram of a typical emulsion is shown in Fig. 189.

They may be handled in the light from a red safelight.

3) Panchromatic emulsions

With still further colour-sensitisation (on the same principle), emulsions can also be made sensitive to red radiations (600-700 mμ).

Fig. 190

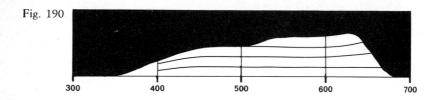

Fig. 190 shows the spectrogram of a typical emulsion. Such emulsions are called "panchromatic" since they are sensitive to all the wavelengths of the visible spectrum. For this reason they must be handled in total darkness.

CHAPTER 2

COLOUR SEPARATION

To be able to reproduce by printing an original in colour with all its variations of hue, saturation and brightness, it is necessary to make use of the technique of "colour separation".

The knowledge and application of this technique is the purpose of this book: we shall thus describe it in detail under the following three headings: 1) The principle of separation. 2) Method of separation. 3) The deficiencies of separation.

1) THE PRINCIPLE OF SELECTION

We already know that *printing inks* are none other than colour pigments and, as such, combine to give different colours including black on the basis of subtractive synthesis.

We also know that the colours used in this synthesis are the primary pigment colours, Cyan, Magenta and Yellow (Figs. 161 and 162).

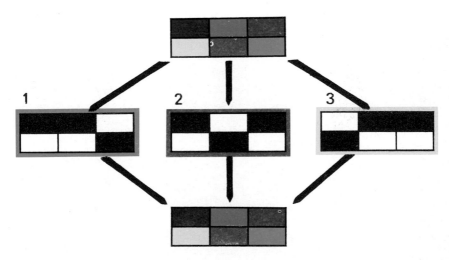

Fig. 191

Thus if we could *break down* the coloured original into three black-and-white photographs (Fig. 191) one of which reproduces the Cyan content of the original, another the Magenta content of the original and a third the Yellow content of the original, *we should be able to build up* the coloured original by printing, one over the other, the design of photograph 1 with Cyan ink, the design of photograph 2 with Magenta ink and the design of photograph 3 with Yellow ink.

2) METHOD OF SEPARATION

This *breaking down* of the original is made possible by the use of colour separation filters (Footnote 1).

Footnote (1) Colour separation filter are of two types: 1) *wide band* for the separation of opaque originals (paintings, etc.); 2) *narrow band* for makings separations from transparent originals (colour transparencies, colour negatives, etc.).
Below we reproduce the spectral absorption curves of the two series of Kodak Wratten filters:
Wide band filters

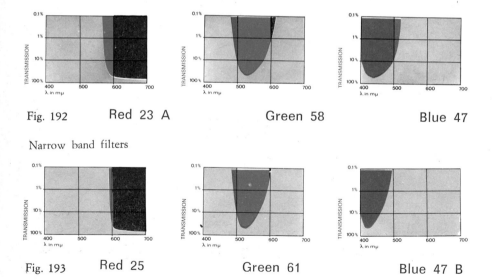

Fig. 192   Red 23 A   Green 58   Blue 47

Narrow band filters

Fig. 193   Red 25   Green 61   Blue 47 B

The equivalent Ferrania filters are: wide band — Red 03, Green 02 and Blue 01; narrow band — Red BS 06, Green BS 07 and Blue BS 08. The equivalent Agfa-Gavaert filters are: wide band — R 599, G 517 and B 488; narrow band — R 619, G 537 and B 479.
When the separations are to be made optically (i.e., not by contact, but with a camera or enlarger) is it necessary to use « optical » filters to avoid loss of image quality. Kodak Wratten filters of this quality are prefixed with PM (Photomechanical) and those of Agfa-Gevaert are followed by the letter C.

If we make an exposure of a coloured original on panchromatic film through a red filter (Fig. 194) (remember that red = magenta + yellow, see Fig. 158), only the red radiation will be transmitted to form the first photograph which will then become the separation negative for printing the Cyan image.

From this we can obtain a screened negative or positive (for letterpress or lithography) or continuous tone positive (for gravure) which will serve for printing the complementary to red, that is, the Cyan ink.

By repeating this operation using a Green filter and then a Blue filter, we shall obtain the separations for printing with the Magenta and Yellow inks respectively.

Finally by printing one over the other, (known as "progressive proofing"), we shall finish with a reproduction of the original.

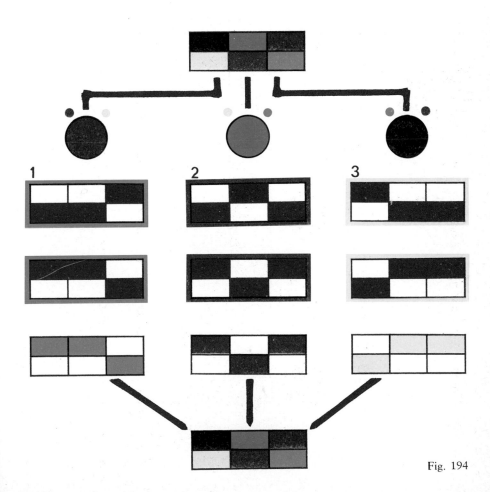

Fig. 194

145

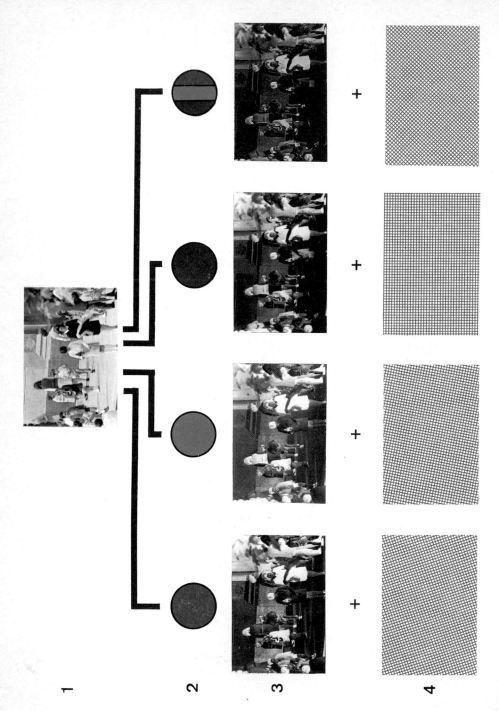

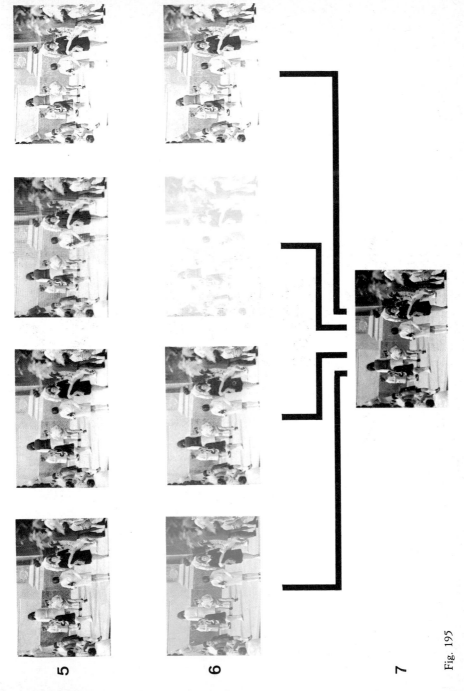

5

6

7

Fig. 195

147

Let us now look at a practical demonstration of colour separation (Fig. 195):

The original (1) (in this case a colour transparency, but the same method would apply to an opaque colour original) is photographed separately through Red, Green and Blue filters (2) on a normal tone panchromatic film. From these exposures we obtain the separation negatives (3) for the Cyan, Magenta and Yellow. By making successive exposures through the three filters, we obtain the negative for the black printing.

From the four separation negatives, we can obtain through the half-tone screen (4) using an orthochromatic lith film, the screened positives (5) for Cyan, Magenta, Yellow and Black.

From these we prepare the lithographic plates for printing the colour progressives (6) and the final reproduction (7).

3) THE DEFICIENCIES OF COLOUR SEPARATION

Fig. 196

In Fig. 196 we repeat, on a slightly larger scale, the original and its reproduction by colour separation printing illustrated in Fig. 195. A glance is sufficient to notice that the printed reproduction does not match up to the quality of the original. It appears reddish (impure hue), the colours are dirty (poor luminosity) and dull (poor saturation).

What is the cause of these imperfections?

We can answer immediately that the cause is chiefly to be found in the *lack of purity in the colours of the printing inks* (Footnote 1), for which

Footnote (1) We have said « chiefly ». In fact a secondary cause is to be found in the separation filters themselves which do not separate the colour perfectly. However, we need not concern ourselves with the filters for two reasons:

1) because the inks are by far the chief culprits;

2) because the « fundamental method » of correcting the separations, which we shall be explaining later (page 166), automatically corrects filter defects as well as impurities of the inks.

reason the simple technique of colour separation so far described is not, by itself, sufficient to assure us of acceptable results.

In the next chapter we shall examine in detail methods for overcoming this deficiency. For the moment, let us look closer at the cause: the defective inks.

A "standard" series of modern three-colour printing inks, namely, Cyan, Magenta and Yellow, generally presents defective absorptions and reflectances which we can consider on the basis of the following average data:

1) The Cyan ink

If the Cyan ink were pure, it would absorb (as we already know) only the Red radiations and would completely reflect the Green and Blue radiations (Fig. 197).

Fig. 197

Fig. 198

Unfortunately the colour pigments available to the ink manufacturers are not pure and the inks they succeed in producing are consequently defective.

In fact, in practice, a Cyan ink has these three defects (Fig. 198):

a) it absorbs a small amount of Blue that should not be absorbed;
b) it absorbs a notable amount of Green that it should not absorb;
c) it reflects a very small quantity of Red that it should not reflect.

2) The Magenta Ink

If the Magenta ink were pure, it would absorb only the Green radiations, and would completely reflect the Red and Blue radiations (Fig. 199).

149

Fig. 199   Fig. 200

However, in practice, a Magenta ink has these three defects (Fig. 200):

a) It absorbs a considerable amount of Blue radiations that it should not absorb;

a) it reflects a small amount of Green radiations which is should not reflect;

c) it absorbs a small amount of Red radiations that it should not absorb.

3) The Yellow ink

If the yellow ink were pure, it would absorb only the Blue radiations and completely reflect the Green and Red radiations (Fig. 201).

Fig. 201   Fig. 202

In practice the Yellow ink has only these two defects (Fig. 202):

a) It reflects a very small amount of Blue radiations which it should not reflect;

b) it absorbs a very small amount of Green radiations which it should not absorb (Footnote 1).

Footnote (1) We may note at once that the defects of the inks which modify the hue are those of excessive absorption rather than excessive reflection. Our chief concern, therefore, as we shall see in the next chapter, is to correct for these unwanted absorptions.

150

A) **Spectrophotometric data**

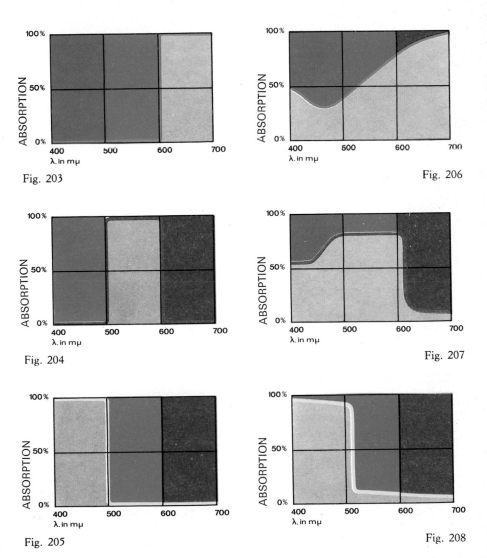

Fig. 203

Fig. 206

Fig. 204

Fig. 207

Fig. 205

Fig. 208

If the three inks, Cyan, Magenta and Yellow, were pure (ideal colours) their "spectral absorption curves" would be those shown in Figs. 203, 204 and 205 (left hand column): The absorbed radiations are shown in grey and those reflected (or transmitted) in their respective colours.

In Figs. 206, 207 and 208 (righthand column) the "spectral absorption curves" of the actual inks are shown. An examination of them will lead to the same conclusions illustrated in Figs. 197 to 202.

B) Densitometric data

If, with a reflection densitometer, we measure the densities of the three standard inks, Cyan, Magenta and Yellow printed on white paper, first through a Red filter, then through a Green Filter and finally through a Blue filter, we shall obtain the following average readings (Fig. 209):

TABLE OF REFLECTION DENSITIES

	READING THROUGH THE FILTER		
COLOUR OF INK:	RED	GREEN	BLUE
CYAN	1.57	0.50	0.20
MAGENTA	0.20	1.38	0.67
YELLOW	0.02	0.10	1.67

Fig. 209

The table confirms what we already know, that is:

1) that the Cyan ink absorbs (besides RED - D = 1.57), a considerable amount of Green (D = 0.50) and a small part of the Blue (D = 0.20),
2) that the Magenta ink absorbs (besides Green - D = 1.38) a great deal of Blue (D = 0.67) and a small part of Red (D = 0.20);
3) that the Yellow ink absorbs (besides Blue - D = 1.67) some green (D = 0.10), but an insignificant amount of red (D = 0.02).

Interpretation of the densitometric data

But what chiefly interests us is the interpretation of the data in the table, so as to know beforehand what modification we shall have to apply to the printing matrices of the three colours Cyan, Magenta and Yellow, to obtain with the inks being tested, a print that faithfully reproduces the original.

Let us assume that:

The Red, Green and Blue *filters* (shown at the head of the table), respectively absorbing the Cyan, Magenta and Yellow radiations, represent in fact these three colours in the pure state (ideal).

The Cyan, Magenta and Yellow *inks* (shown on the left of the table), represent instead these three colours in the impure state (actual).

In addition, let us call the highest density H, the medium density M, and the lowest density L.

Thus applying it to our data, for Magenta ink: H = 1.38; M = 0.67; L = 0.20 and for the Red filter (pure Cyan): H = 1.57; M = 0.02; L = 0.02.

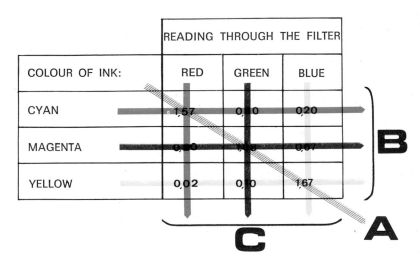

Fig. 210

And here is how the table is interpreted (Fig. 210):

1) **Line A, or of the ink colours**

The "boxes" crossed by the diagonal line A register *the Densities (H) of the pure colour contained in each of that colour.*

2) **Line B, or of the inks**

The boxes crossed by the horizontal line B register, for each colour, *the Densities (H, M, and L) of the pure colours (H) or impure (M and L) contained in it.*

3) **The line C, or of the colours (complementaries of the filters)**

The boxes crossed by the vertical lines C register, for each colour, *the Densities (H, M and L) that indicate how much of this colour is contained in the various inks.*

In particular, the boxes M and L of line C tell us *how much of that colour is unduly contained* in the relative inks (footnote 1), and thus by how much it must be corrected (masked) at the separation stage (see later in chapter 3).

GATF characteristics

Although it is not strictly necessary to the treatment of our present subject, it is useful to make some mention of the GATF standards (Graphic Arts Technical Foundation) that characterise a printing colour.

They are useful to the photomechanical technician in evaluating the inks by simple reference to the "GATF characteristics" which manufacturers now allot to their products, especially to the series of inks for three-colour printing (Cyan, Magenta and Yellow).

The experts of GATF have devised several simple formulae to determine, on densitometric data — four characteristics (two positive and two negative) of printing inks:

1st The Ink Strength
2nd The Visual Colour Efficiency
3rd The Hue Error
4th The Greyness Error

1) The first characteristic (positive) of a coloured ink is the *Strength,* that is, its printing power. This is given, for each ink, by the Density H (line A). In the case of the inks in the table (Fig. 210), the Cyan ink has a strength of 1.57, the Magenta of 1.38 and the Yellow of 1.67. Knowing the strength of the inks we can anticipate that in superimposing, for example, a Yellow ink of strength 1.67 to a Magenta ink of strength 1.38, the resulting Red will be rather yellowish, since in this case the Yellow is stronger than the Magenta.

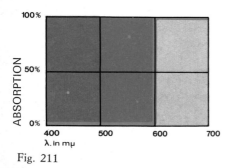

Fig. 211

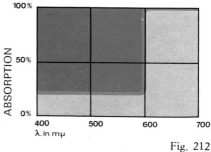

Fig. 212

Footnote (1) The Density M represents the « major » impurity; the Density L the « Minor » impurity.

154

2) The second characteristic (positive) of an ink is the *Visual Colour Efficiency,* that is, its capacity of "reflecting to the maximum" the luminous radiations which it should reflect, without absorption.

Fig. 211 show the spectral absorption curve of a theoretical Cyan ink having the maximum efficiency.

It is then said to have an Efficiency of 1 (or 100%).

Fig. 212 shows the absorption of a theoretical Cyan ink with an Efficiency of 80%.

When there are unwanted absorptions in an actual ink, the Efficiency will be given by the formula:

$$\text{Efficiency} \quad \frac{L + M}{2 \times H}$$

where 1 represents maximum efficiency (100%) and $\dfrac{L \mid M}{2 \times H}$ the percentage of unwanted absorption.

The Magenta given in the table (Fig. 210) will thus have an Efficiency of: Efficiency of Magenta $= 1 - \dfrac{0.20 + 0.67}{2 \times 1.38} = 1 - \dfrac{0.87}{2.76} = 1 - 0.31$ $= 0.69$ (or 69%).

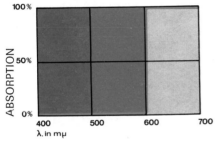

Fig. 213

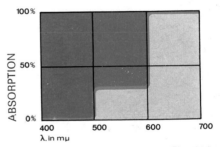

Fig. 214

3) The third characteristic (this time negative) of the ink is the *Hue Error,* that is, that defect which arises from a unbalanced reflection of the colours which should be reflected in equal amounts.

Fig. 213 shows the spectral absorption curve of a theoretical Cyan ink with a hue error (in this case in terms of the Blue, so that the colour will appear more Blue) of 30% (Footnote 1).

Footnote (1) We can note the difference between the « Efficiency » and « Hue error » by comparing the curves in Figs. 212 and 214. The Cyan in Fig. 212 has an Efficiency of 80% but no error in hue (the reflection of Blue and Green are balanced); that in Fig. 214 has a greater Efficiency (85%: in fact the unwanted absorption is some 30% only for the Green, that is, some 15% of the total reflected radiations) but a hue error in terms of Blue of 30%.

The hue error of an actual ink is given by the formula:

$$\text{Hue Error} = \frac{M - L}{H - L}$$

The Magenta in our table (Fig. 210) will thus have the following hue error:

$$\text{Hue Error of Magenta} = \frac{0.67 - 0.20}{1.38 - 0.20} = \frac{0.47}{1.18} = 0.40 \ (\text{or } 40\%).$$

4) The fourth characteristic (negative) of an ink is the *Greyness Error,* that is, the percentage of grey contained in the ink.

This in given by the formula:

$$\text{Greyness Error} = \frac{L}{H}$$

The Magenta ink in the table (Fig. 210) will thus have the following greyness error:

$$\text{Greyness Error of Magenta} = \frac{0.20}{1.38} = 0.14 \ ((\text{or } 14\%).$$

The Graphic Arts Technical Foundation has also prepared *three graphic representations* of the characteristics of inks:

the colour circle, the colour triangle and the colour hexagon.

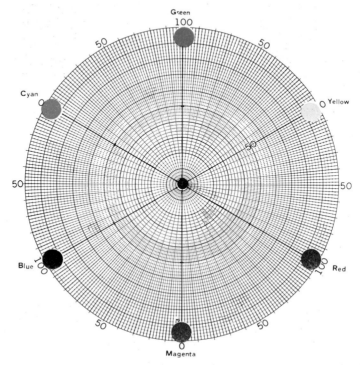

Fig. 215

156

The one of chief interest to us is the Colour Circle (Fig. 215) since it visualises the two characteristics which most help us to know for the correction of the separations: the Hue Error and the Greyness Error.

1) The colour circle is divided by radii into six equal sections, each of which is subdivided into ten small sections.

The three points on the circumference (Fig. 215) where the radii intersect (at 10, 2 and 6 o'clock) are allocated to the subtractive primaries, Cyan, Magenta and Yellow, which carry the figure 0 indicating zero in the "greyness error".

The three points (at 12, 4 and 8 o'clock) are allocated to Green (Yellow and Cyan printed together), Red (Magenta and Yellow printed together) and Blue (Cyan and Magenta printed together). They carry the number 100, indicating a 100% error of hue: Red in fact is a Magenta whose error of hue in terms of Yellow, is 100%.

The subdivisions of the main sections indicate percentages from 10 to 100% in the hue error.

2) The ten main concentric circumferences inside the circle divide the radii into ten equal parts which indicate the percentage "greyness error" (i.e the percentage of grey). On the outer circumference the greyness error is equal to zero: at the fifth division to 50% and at the centre to 100%, which is equal to black.

3) The characteristics of a series of inks, Cyan, Magenta and Yellow can be visualised in the Colour Circle in the following way:

a) the reflection densities of the three inks are measured and registered in the table of densities. Let us suppose we have obtained the data shown in Fig. 209, repeated here in Fig. 216.

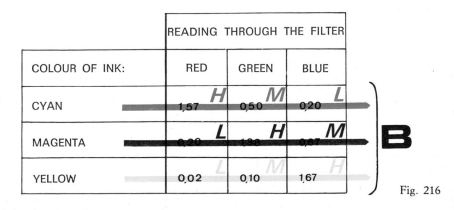

COLOUR OF INK:	READING THROUGH THE FILTER		
	RED	GREEN	BLUE
CYAN	1,57	0,50	0,20
MAGENTA	0,29	1,38	0,67
YELLOW	0,02	0,10	1,67

Fig. 216

157

b) From the formulae explained on page 156, the hue error ($\dfrac{M-L}{H-L}$) and the greyness error ($\dfrac{L}{H}$) is worked out for each of the three colours:

Hue Error: Cyan   = 22 %
                Magenta   = 40 %
                Yellow   = 4.8%

Greyness Error: Cyan   = 12.7%
                Magenta   = 14 %
                Yellow   = 1.2%

c) The values so obtained are plotted on the colour circle as follows:

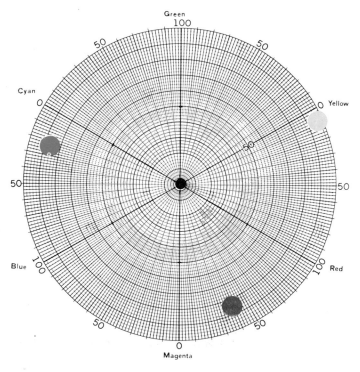

Fig. 217

The *hue error* will place the point representative of the colour on the radius corresponding to its percentage in the direction of the colour which corresponds to the major impurity (Footnote 1).

The greyness error will position the point towards the centre of the circle on the circumference corresponding to the percentage error.

Conclusion

At the end of this analysis of the defects of printing inks — the chief cause of the poor results in three-colour reproduction — it behoves us to remember the following very important conclusions which apply to average printing inks at present in use:

1) *The Cyan ink* by unduly absorbing Green (to a large amount) and Blue (in small amount) behaves *as if it were degraded* by a large amount of Magenta ink (which absorbs Green) and a small amount of Yellow ink (which absorbs blue).

The Magenta ink is thus the major impurity of Cyan and the Yellow ink the "minor" impurity.

a   b

Fig. 218

We can thus represent an ideal Cyan ink (pure) as Fig. 218a, while that of the actual ink (impure) as Fig. 218b.

2) *The Magenta ink,* by unduly absorbing Blue (in large amount) and Red (in small amount) behaves *as if is were degraded* by a large amount of Yellow ink (which absorbs Blue) and a small amount of Cyan ink (which absorbs Red).

The Yellow ink is the "major" impurity and the Cyan ink the "minor" impurity.

a   b

Fig. 219

Footnote (1) It can thus be noted that the GATF circle only registers the major impurity of the « Hue error ».

159

We can thus represent an ideal Magenta ink (pure) as Fig. 219a, while that of the actual Magenta ink (impure) as Fig. 219b.

3) *The Yellow ink* in absorbing only a very small amount of Green, behaves as if it were degraded with a very small amount of Magenta ink (which absorbs Green).

a

b _____

Fig. 220

We can thus represent an ideal Yellow ink (pure) as Fig. 220a, while the actual Yellow ink (impure) as Fig. 220b.

4) *The percentages* of these impurities, whether large or small, are given from "spectral absorption curves" and the "table of densities".

The GATF "hue error", as we have said, gives only the "major" impurity.

CHAPTER 3

COLOUR CORRECTION: MASKING

The separation of colours by means of filters is, as we have seen, insufficient to assure us of a faithful reproduction of the original because the printing inks at our disposal are not as pure as they should be. It is, therefore, necessary to integrate the process of separation with one of "correction" to allow for the impurities of the inks. We shall deal with this problem in three sections: 1) The principle of correction; 2) The fundamental method; 3) Other methods.

1) **THE PRINCIPLE OF COLOUR CORRECTION**

From the conclusions reached at the end of the preceding chapter (page 159) we have grasped that the impurities contained in the three primary inks of Cyan, Magenta and Yellow may be considered as "contaminations" caused by inks of other colour.

The two inks most at fault are (Fig. 221):

Fig. 221

1) *Cyan,* contaminated with Magenta (major impurity) and with Yellow (minor impurity)
2) *Magenta,* contaminated with Yellow (Major impurity) and with Cyan (minor impurity).

Yellow is practically a pure colour.

Ignoring for the moment the major impurity of Magenta and considering only the major impurity of Cyan, let us suppose that we wish to obtain a Blue by overprinting Cyan and Magenta.

Their combined images will give a violet-blue (Fig. 222):

Fig. 222

while if the Cyan had been pure, the result would have been a pure Blue (Fig. 223):

Fig. 223

We might now ask ourselves, how we can obtain a pure Blue with a Cyan ink contaminated with Magenta?

Since the Magenta *in excess* is inseparably a part of the Cyan ink, there remains only one solution open to us: to diminish the amount of Magenta ink in proportion in which it is already present in the Cyan ink (Fig. 224):

Fig. 224

This is the principle of colour correction which, generalised, we shall apply to achieve all the necessary corrections.

A) Correction masks

It now remains to ask: how can we reduce the amount of Magenta in the zones in which the Cyan ink is superimposed?

The answer is, by means of correction masks (Footnote 1).

To understand how this comes about, we must refer again to Fig. 194 which we have reproduced in Fig. 225 only as it affects the reproduction of an original Blue.

From the original (Blue) we obtain, through the Red and Green Filters, the separation negatives for printing Cyan and Magenta.

Footnote (1) In photography, the « mask » is a photographic image which is super-imposed (like a mask over a face) over another photographic image to increase the density in certain zones.

Various uses can be made of it: there are masks for reducing (negative + positive) or increasing (negative + negative or positive + positive) the scale of density; « high-light » masks; colour correction masks, etc.

However, when we refer, as in this Chapter, to « masks » without qualification, we are referring to colour correction masks.

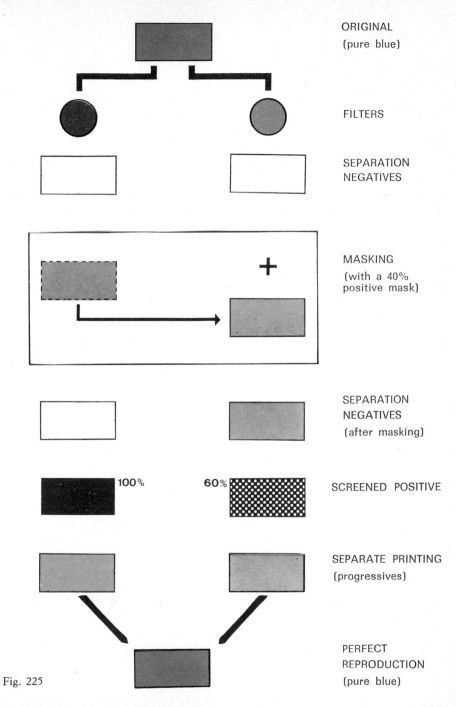

ORIGINAL
(pure blue)

FILTERS

SEPARATION
NEGATIVES

MASKING
(with a 40%
positive mask)

+

SEPARATION
NEGATIVES
(after masking)

100%   60%   SCREENED POSITIVE

SEPARATE PRINTING
(progressives)

PERFECT
REPRODUCTION
(pure blue)

Fig. 225

163

The masking is done at this stage: if the impurity of Magenta contained in the Cyan ink is 40% of the Magenta ink, we shall prepare a positive mask of the Cyan having a density of 40% of the Magenta and superimpose it on the Magenta separation negative which will become denser by 40% in those zones in which the Cyan and Magenta inks will be printed one over the other to give Blue. In this way the screened positive of the Magenta will diminish (in the zones overprinted with Cyan) the dot percentage by 40%, that is, it will print 40% less Magenta, thus giving a correct reproduction of the original Blue.

Hence we may thus conclude:

The correction of the Magenta impurity contained in the Cyan ink is achieved by *masking* the Magenta separation negative with a positive obtained from the Cyan separation negative, which has a density (usually spoken of as "strength") directly proportional to the impurity to be corrected.

This, obviously, applies also to the correction of Cyan and Yellow impurities.

B) The "strength" of the mask

The calculation of the density percentage or "strength" of the mask must be made on the basis of the spectrophotometric data (spectral absorption curves) or on the basis of densitometric data (table of densities). Usually, we make use of the latter (Fig. 226).

COLOUR OF INK:	READING THROUGH THE FILTER		
	RED	GREEN	BLUE
CYAN	1,57	0,50 *M*	0,20
MAGENTA	0,20	1,38 *H*	0,67
YELLOW	0,02	0,10 *L*	1,67

C

Fig. 226

The unwanted colour present in each ink is given by the density M (major impurity) and L (minor impurity) of the lines "C" of Magenta: *the colour Magenta* (separated by the Green filter) is contained in the *Magenta ink* as a density of 1.38 (H), but in also contained as an impurity in the *Cyan ink* as a density of 0.50 (M) and in the *Yellow ink* as a Density of 0.10 (L).

Since 0.50 is 36% of 1.38 (Footnote 1), the Cyan ink will contribute an unwanted 36% of Magenta *in the zones which have received Magenta ink.* It follows that the mask to reduce the amount of Magenta ink in these zones must have a density to 36% *of the density of the Magenta separation negative.*

If therefore the Magenta separation negative has, for example, a density scale of 1.20, the correction mask with a strength of 36% must have a density scale of 0.45, namely 36% of 1.20 (Footnote 2).

Since 0.10 is 7% of 1.38, the *Yellow ink* will contribute an unwanted 7% of Magenta in the zones which have received Magenta ink. It follows that the mask to reduce the amount of Magenta ink in the said zones must have a density equal to 7% *of the density of the Magenta separation negative.* If therefore the Magenta separation negative has a density scale of 1.20, the correction mask must have a density scale of 0.084 which is 7% of 1.20.

The same calculation is repeated for the other lines "C" of the colours Cyan and Yellow.

However in practice, the densities of the correction masks in respect to the densities of the image that must be corrected, approximate to the following average percentages (Fig. 227) which may be adopted with excellent results with modern inks of good quality:

Footnote (1) It is useful to recall what has been said on page 52. Percentage is nothing more than a proportion which can be stated thus:

$$1.38 : 100 = 0.50 : x \qquad \text{from which}$$

$$x = \frac{100 \times 0.50}{1.38} = \frac{50}{1.38} = 36$$

Footnote (2) In fact the percentage of the « strength » of the mask is given by the following formula:

$$\% \text{ strength} = \frac{\text{D. Scale of original}}{\text{D. Scale of mask}} \times 100$$

where in this case the original is represented by the separation negative.

The separation to be printed with ink...	Requires a mask made from the separation of the...	with a density (strength) of %... in respect of the density of the separation	Impurity to be corrected:
Cyan	Magenta	15 %	MINOR
	Yellow	2 %	NEGLIGIBLE
Magenta	Cyan	40 %	MAJOR
	Yellow	5 %	NEGLIGIBLE
Yellow	Magenta	40 %	MAJOR
	Cyan	15 %	MINOR

Fig. 227

2) THE FUNDAMENTAL METHOD

Among all the methods in use today for the correct separation of the colours, which for its completeness (in fact it embodies all the advantages of other methods), for its practicability in terms of control and for the perfection that can be achieved in the results can well be considered the fundamental method.

We shall describe two versions of it: the first which corrects only the major impurities; the second (an extension of the first) which also corrects the minor impurities. Finally, we shall illustrate the almost unlimited possibilities offered by this method for special effects.

The method applies equally to opaque and transparent originals.

A) Separation with correction for the major impurities

It is accomplished in 5 stages (Fig. 228):

Stage 1 - *Density Scale.* If the original is a transparency (Footnote 1) it may be necessary to reduce the density scale by means of a negative mask

---

Footnote (1) The density scales of transparencies may be classed as *low* (less than 2), *medium* (from 2 to 2.5) and *high* (more than 2.5). If there are many transparencies to be dealt with at one time, they should be grouped according to their density scales.

However, the proportional mask, being negative, when combined with the transparency will automatically level out the differences between the various transparencies.

166

(Footnote 1). This is made from the transparency on a suitable panchromatic film (Footnote 2) with white light.

This stage is not necessary for opaque originals which generally have a normal density scale.

Stage 2 - *Separation.* These are made through the original plus the mask (Footnote 3) with Red, Green and Blue filters to give the Cyan, Magenta and Yellow separations: by making successives exposures through all three filters to obtain the negative for the black printing.

Stage 3 - *Preparation of pre-masks.* These are made by contact from the Magenta and Yellow separation negatives on an ortho or blue-sensitive "tone" film so that the positive images are such that, when combined with their negatives, the image completely disappears giving place to an area of uniform density (100% mask).

Pre-masks are the foundation of what we have called the "Fundamental method"; we shall deal with them fully in the "NB" on page 170.

Stage 4 - *Preparation of the 40% correction masks.* The pre-mask of the Magenta separation (Footnote 4) is combined with the Cyan separation negative in order to obtain the 40% Cyan correction mask, the exposure being made on ortho or blue-sensitive film. The Yellow pre-mask is combined with the Magenta separation negative to obtain the 40% Magenta correction mask.

Stage 5 - *Screening.* The Cyan correction mask is combined with the Magenta separation negative, and that of Magenta with the Yellow separation negative. We then proceed to the screening of the four separation negatives.

Footnote (1) The negative masks (in relation to the original) are obtained directly from the original (positive); the positive masks, on the other hand, are obtained from the separation negatives.

Footnote (2) Excellent results, as well as an initial correction of the colour, can be obtained by using for this mask the Agfa-Gevaert « Verimask » film or that of Du Pont marketed under the name of « Neomask » film. The same initial correction of the colour can be obtained by using an ordinary « tone » panchromatic film to make the contrast reducing mask referred to in Stage 1, giving a main exposure in white light followed by a brief exposure in red light (the Cyan separation filter).

Footnote (3) For the rapid and accurate registration of the masks (which are used frequently in this method), it is important to perforate both the original (or the support to which it is taped) as well as all the films used for masking before the exposures are made. The perforations given by an ordinary office perforator are very satisfactory, when used with the registration pins which are available commercially.

Footnote (4) The masks are always specified by the colour of the separation negative from which they have been made. This, for example, is called the « Magenta pre-mask » since it has been obtained from the Magenta separation negative.

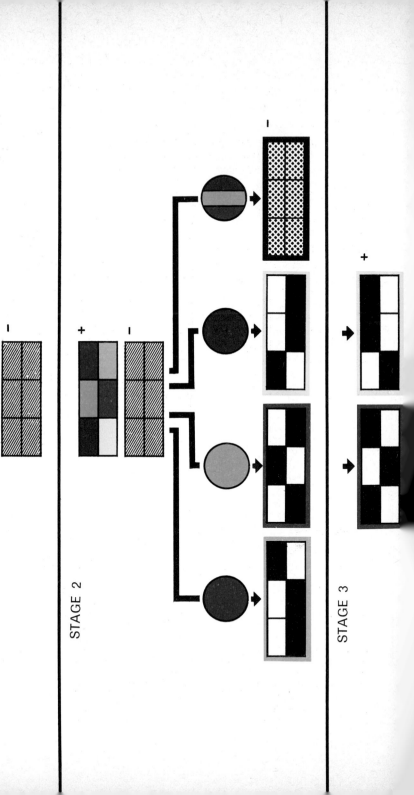

STAGE 1

STAGE 2

STAGE 3

168

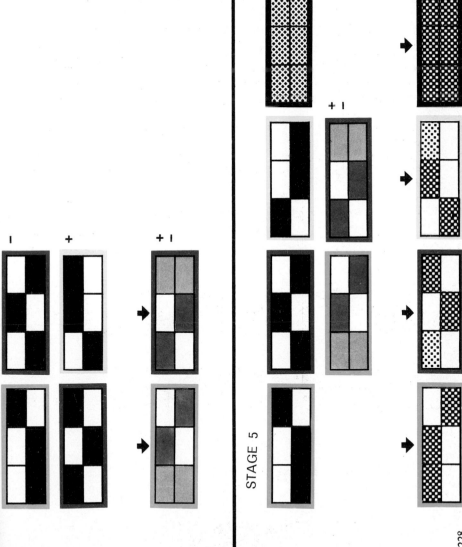

STAGE 5

Fig. 228

N.B. - THE PRE-MASKS

We can now more readily understand the function of the pre-mask: if it had not been there, the 40% correction Cyan mask would have reduced the printing strength of the Magenta in Cyan, Green and Blue in proportion to the density of the Cyan.

This masking is *theoretically* exact for the reason we have already given (page 162 and Fig. 225) but *in practice* it does not give fully satisfactory results.

In fact, in practice, various factors, such as poor whiteness of the paper, the admixture of colours which arises inevitably during the printing stages of a multi-colour printing machine (for example, a roller printing one colour over another colour which is not yet dry), but above all the "subjective" need of the human eye to see the separation of hues exaggerated, make it is necessary to take the separation of the colours *beyond* what is theoretically accurate.

In fact, these practical results can only be fully achieved by making use of the pre-masks.

To understand this and the other advantages they can contribute, let us compare two correction masks obtained one *without* pre-masking and the other *with*.

a) Positive mask of 40% correction obtained with pre-mask

For example, the Cyan corrective mask is taken from the separation negative for Cyan and then bound up with the separation negative from Magenta (Fig. 229).

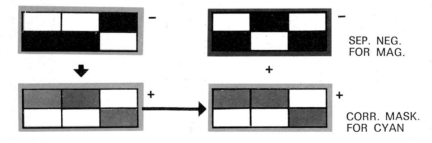

SEP. NEG.
FOR MAG.

CORR. MASK.
FOR CYAN

Fig. 229

It can be represented schematically as in Fig. 230:

Fig. 230

170

— In zones (A) where there is no Cyan (Magenta, Red and Yellow zones of the original), there will be *no modification* to the printing strength of the Magenta.

— In zones (B) where there is Cyan (Cyan, Green and Blue zones of the original) there will be *the maximum reduction* (40%) of the printing strength of the Magenta.

The diagram - Fig. 231 - gives the theoretical curve of this masking.

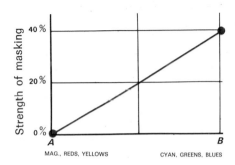

Fig. 231

b) **Positive-negative masking at 40% obtained with pre-mask**

This is obtained, as we already know, according to the scheme shown in Fig. 232.

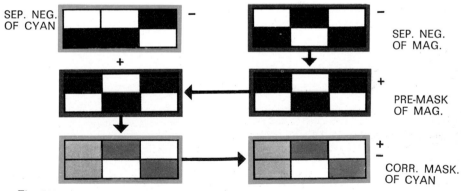

Fig. 232

It can be schematically represented as in Fig. 233.

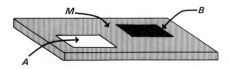

Fig. 233

171

— In zones (M) where there is no Cyan or Magenta (Yellow zones of the original) or where Cyan and Magenta are of equal strength (Blue zones of the original), there is *no modification* of the printing strength of the Magenta. (Equals layer of mean grey, 20%: adds density but does not correct).

— In zones (A) where there is Magenta but not Cyan (Magenta and Reds of the original) there is a *maximum raising* of the printing strength of Magenta (equals transparent layer, 0%).

— In zones (B) in which there is Cyan but not Magenta (Cyans and Greens of the original), there is the *maximum reduction* of the printing strength of the Magenta (equals dark layer, 40%).

Fig. 234 gives the theoretical curve of this masking:

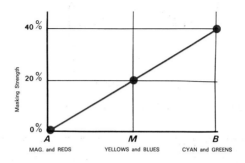

Fig. 234

Here are the principal advantages of this kind of masking:

1) *Emphasis of the separation,* without disturbing the chromatic balance. In fact, the pre-mask "preserves" from blackening only and all those zones of the correction mask in which the colour it prints *must* print (in the example above the Magenta and Reds of the original).

2) The use of the pre-mask *gives rise to a correction mask with positive-negative characteristics* (in our example, positive of Cyan, negative of Magenta), which have the great advantage of not modifying the Scale of the separation negative.

3) The particular "three layer" configuration (A, B and M) of the correction masks obtained by means of the pre-masks, *automatically corrects also the absorption defects of the filters* (see footnote 1, page 148), especially those of the Green and Blue. This can be readily understood by referring to Fig. 235 where unwanted density caused by the imperfect separation of the filters is virtually cancelled out by the layer M (intermediate) of the mask.

172

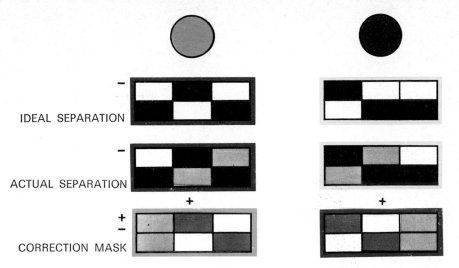

IDEAL SEPARATION

ACTUAL SEPARATION

CORRECTION MASK

Fig. 235

Example of separation

Because the method we have described is already recognized as unsurpassed for making separations from opaque originals, we have chosen to give an example of the procedure for a transparency (Fig. 237) to demonstrate its value in this case also.

By way of interest, we shall give details of the materials used, method of exposure and development for each stage (Footnote 1).

Stage 1 - Film; soft gradation panchromatic.
  Contact exposure for 2 sec. in tungsten light.
  Development in MQ developer for 2 min.
  Density scale of transparency = 2.4; Scale of mask = 0.9;
  Scale of transparency + mask = 1.50.

Footnote (1) We give the « time-gamma » curves for a soft gradation film exposed through the three separation filters, Red, Green and Blue.

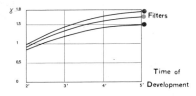

Fig. 236

On the basis of these data, by means of a test, we can construct the time-gamma curve with the apparatus and materials (films, filters, developers) used in our own laboratory which will serve us for all future work.

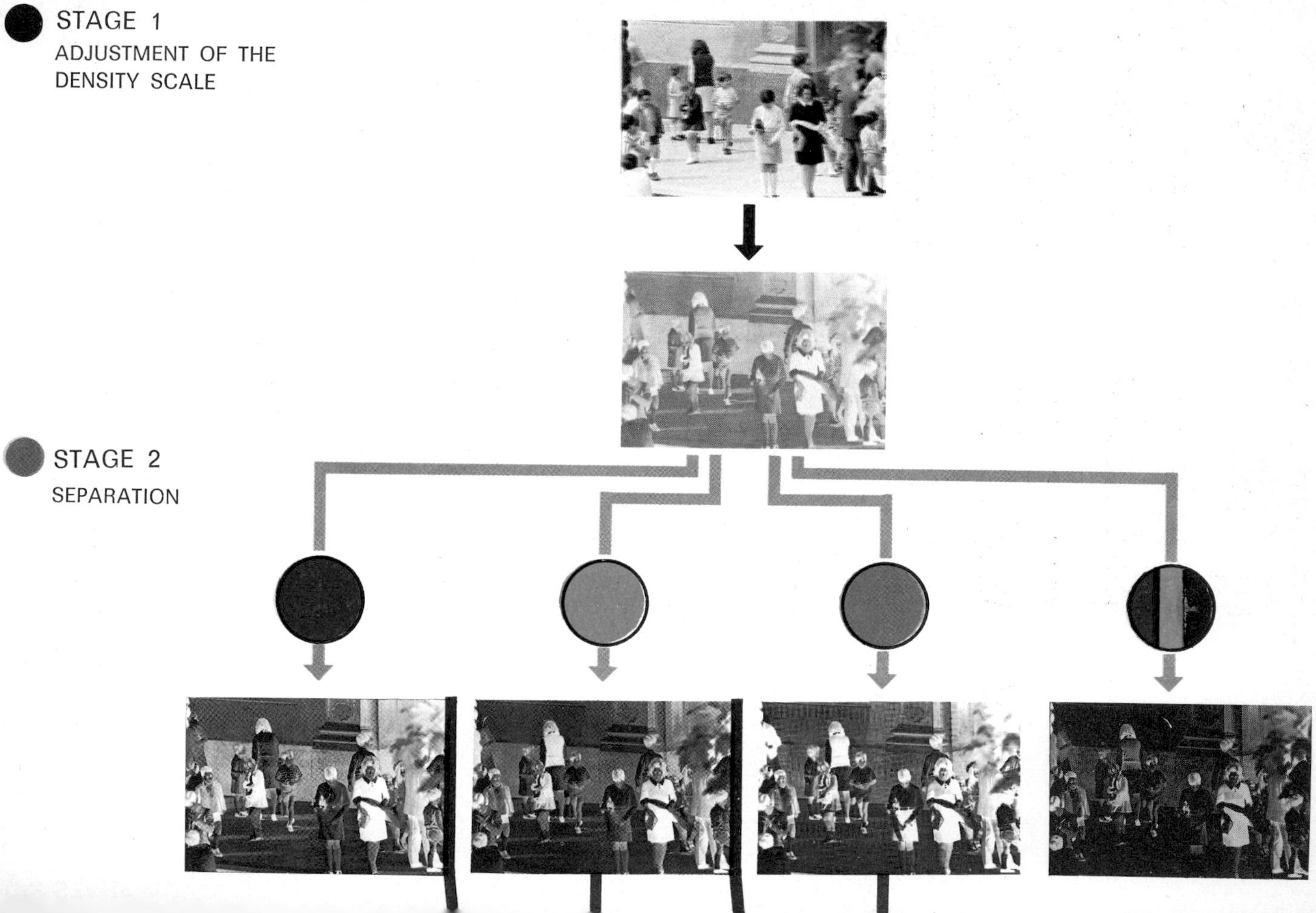

STAGE 1
ADJUSTMENT OF THE
DENSITY SCALE

STAGE 2
SEPARATION

STAGE 3
100%
PRE-MASK

STAGE 4
40%
CORRECTION
MASK

STAGE 5
SCREENING

Fig. 237

175

Stage 2 - Film; medium tone panchromatic for making separation negatives.
   Contact exposure with tungsten light:
   with Red filter for 3 sec., developed for 3 min. in MQ developer
   (Footnote 1)
   with Green filter for 4 sec., developed for 3½ min. in MQ dev.
   with Blue filter for 7 sec., developed for 4 min. in MQ dev.
           Green for 2 sec.
   with   Red for 2 sec.   developed for 4 min. in MQ dev.
           Blue for 4 sec.

Stage 3 - Film; tone orthochromatic of medium contrast.
   Contact exposure with tungsten light for 1 sec.
   Development in MQ developer for 3 min.

Stage 4 - Film; tone orthochromatic of high contrast.
   Contact exposure with tungsten light for 1 sec. with a diffusing
   sheet between film and separations.
   Development in MQ developer for 3 min.

Stage 5 - Film; lith orthochromatic of high contrast.
   Exposure with precalculation of the basic contrast through a Magenta screen.
   Development in paraformaldehyde developer.

N.B. - HIGHLIGHT MASKS

Let us begin by saying that when we speak of "lighter tones" and "shadows" we refer always to the *original positive.*

The *lighter tones* of the original are the clearer parts (more transparent in the case of a transparency) where it is still possible to see some detail or colour even though faint.

The *shadows* of the original are the darker parts (less transparent).

The *highlights* of the original are those very clear points (transparent) which appear white.

Such, for examples are the reflections from a vase or well-polished apple, the reflections from water, from polished metal, etc.

When it is required to emphasise these highlights by eliminating even the 5% dots, it is necessary to make use of a highlight mask.

Footnote (1) See footnote 1 on page 173.

This mask (negative) is obtained directly from the original positive using a lith film of high contrast. Correctly made, it is completely clear except in the highlight areas which must have a density of about 0.5.

The mask is placed in register with the separation negatives at the screening stage.

B) **Separations with correction for major and minor impurities**

If we wish to obtain a perfect correction of the colours we can take the process further to include also the minor impurities. In this case, stages 1 and 2 are unchanged, stage 3 modified, stage 4 unchanged, a new stage 5 is added (preparation of the 15% correction masks) and stage 6 (previously 5) modified: see below (Fig. 238):

Stage 3 - *Preparation of 100% pre-mask.* In this case, three instead of two masks are made, one from each of the separation negatives.

Stage 4 - see Stage 4 in Fig. 228.

Stage 5 - *Preparation of 15% correction masks.* This is carried out as follows:
The pre-mask for Yellow is used with the Cyan separation negative to obtain the 15% Cyan correction mask.
The Cyan pre-mask is used with the Magenta separation negative to obtain the 15% Magenta correction mask.

Stage 6 - *Screening.* The separation negatives are masked as follows:
The Cyan separation negative is masked with the 15% Magenta correction mask.
The Magenta separation negative is masked with the 40% Cyan correction mask.
The Yellow separation negative is masked with both the 40% Magenta and 15% Cyan corrections masks.

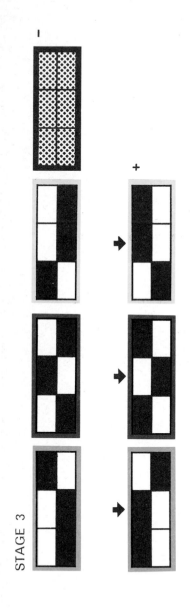

STAGE 3

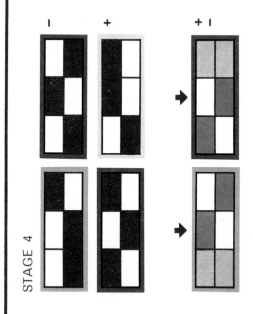

STAGE 4

178

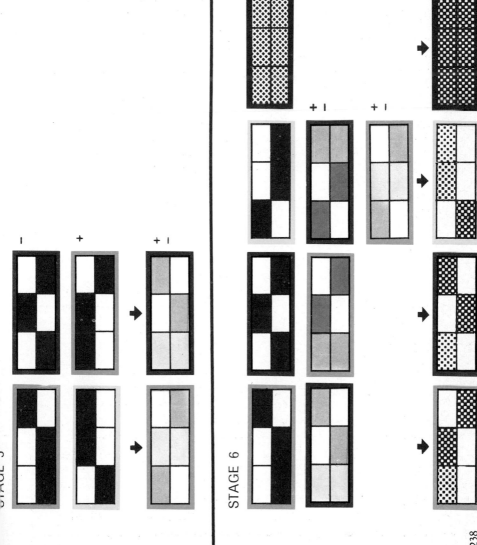

STAGE 6

Fig. 238

179

C) Special effects

The technique so far described for the correction of colours can also be used for obtaining special effects in the reproduction of colour originals.

Fig. 239 to 242 illustrate the four principal combinations between *separation negatives and 100% pre-masks* and the resulting *correction masks*. The latter may have a "strength" of from 10% to 90% according to the effects that we wish to obtain and must be used in register with the separation negative from which the pre-mask has been made.

The four positives (screened or continuous tone) obtained are dealt with in the order: Cyan 1, Magenta 2, Yellow 3, Cyan 4.

These will replace the "normal" screened positives with the purpose of obtaining the special effect desired.

In Fig. 243 technical data concerning the original transparency is given, and in Fig. 244 the reproduction obtained with the correction of the major impurities.

In Figs. 245 to 248 the original is repoduced using only *one* "special effects" screened positive; in Figs. 249 to 251 with *two* such positives; in Fig. 252 with *all three* positives made for special effects.

The following table summarisees the combinations illustrated in Figs. 244 to 252.

	CYAN	MAGENTA	YELLOW	BLACK
1	1	NORMAL	NORMAL	NORMAL
2	NORMAL	2	NORMAL	NORMAL
3	NORMAL	NORMAL	3	NORMAL
4	4	NORMAL	NORMAL	NORMAL
5	NORMAL	2	3	NORMAL
6	1	NORMAL	3	NORMAL
7	1	2	NORMAL	NORMAL
8	1	2	3	NORMAL

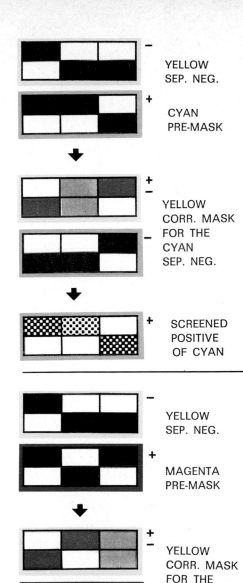

COMBINATION 1

Yellow Neg. + Pre-mask from Cyan

Placed in register with Cyan separation negative to obtain screened positive "CYAN 1".

Special effects

Reduces the Cyan printing ink in the zones printed with Yellow ink. That is:
1) The Yellows will be cleaner (less greenish)
2) The Greens will become more Yellow
3) The Browns become more Orange.

YELLOW SEP. NEG.

CYAN PRE-MASK

YELLOW CORR. MASK FOR THE CYAN SEP. NEG.

SCREENED POSITIVE OF CYAN

Fig. 239

COMBINATION 2

Yellow Neg. + Pre-mask of Magenta

Placed in register with the Magenta separation negative to obtain the screened positive "MAGENTA 2".

Special effects

Reduces the Magenta printing ink in the zones printed with Yellow. That is:
1) The Yellows are cleaner (less Orange)
2) The Reds will become more Yellow
3) The Browns will become more greenish.

YELLOW SEP. NEG.

MAGENTA PRE-MASK

YELLOW CORR. MASK FOR THE MAGENTA SEP. NEG.

SCREENED POSITIVE OF MAGENTA

Fig. 240

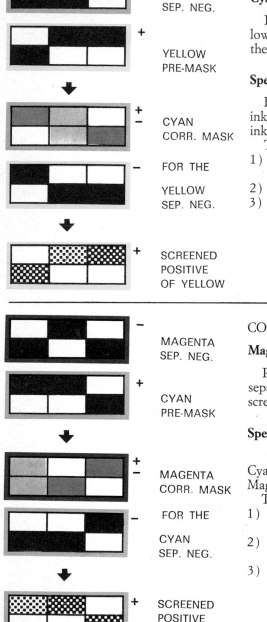

CYAN
SEP. NEG.

YELLOW
PRE-MASK

CYAN
CORR. MASK

FOR THE

YELLOW
SEP. NEG.

SCREENED
POSITIVE
OF YELLOW

COMBINATION 3

Cyan Neg. + Yellow pre-mask

Placed in register with the Yellow separation negative to obtain the screen positive "YELLOW 3".

Special effects

Reduces the printing of Yellow ink in the zones printed with Cyan ink.
That is:
1) The Cyans become cleaner (less greenish)
2) The Greens become more Cyan
3) The Browns become more violet.

Fig. 241

MAGENTA
SEP. NEG.

CYAN
PRE-MASK

MAGENTA
CORR. MASK

FOR THE

CYAN
SEP. NEG.

SCREENED
POSITIVE
OF CYAN

COMBINATION 4

Magenta Neg. + Cyan Pre-mask

Placed in register with the Cyan separation negative to obtain the screened positive "CYAN 4".

Special effects

Reduces the printing of the Cyan ink in the zones printed with Magenta ink.
That is:
1) The Magentas become cleaner (less violet)
2) The Blues become more Magenta
3) The Browns become more Orange.

Fig. 242

182

Original (Fig. 243):
Ektachrome transparency
Rollei 35 with Tessar f/3.5 40 mm. lens
Exposure 1/250 sec. at f/11

Fig. 243

Reproduction (Fig. 244) carried out with the fundamental method, with major impurities corrected only.

To demonstrate the extraordinary versatility of the masks for special effects, we have deliberately chosen a transparency containing mixed colours. For example, the colour of the water contains, besides Cyan, a large amount of Yellow and Magenta, the earth a brown colour (Yellow, Magenta and Cyan) and the colour of the vegetation, which besides Cyan and Yellow, contains a good deal of Magenta.

Fig. 244

Fig. 245 - *Magenta and Yellow normal: Cyan 1*

Fig. 246 - *Cyan and Yellow normal: Magenta 2*

Fig. 247 - *Cyan and Magenta normal: Yellow 3*

Fig. 248 - *Magenta and Yellow normal: Cyan 4*

185

Fig. 249 - *Cyan normal: Magenta 2: Yellow 3*

Fig. 250 - *Magenta normal: Cyan 1: Yellow 3*

Fig. 251 - *Yellow normal: Cyan 1: Magenta 2*

Fig. 252 - *Cyan 1: Magenta 2: Yellow 3*

At the conclusion of this description of the fundamental method, we wish to restate that it is;

1) *The most perfect*: no other method reaches the same high level of correction.
2) *The most versatile*: no other method offers such possibilities for photographic modification.
3) *The most economic*: because the separations and masking is done nearly always by contact the use of sensitised material is limited. In addition, the material is that for black-and-white photography and is thus less expensive.
4) *The most rapid*: separations by contact from transparencies can usually be made of several subjects at the same time: in addition, this method almost totally eliminates the need for manual retouching.

3) OTHER METHODS

Methods for correcting colours by means of masks are of three kinds: with negative grey masks, with positive grey masks, with negative colours masks.

A) Method of correction with negative grey masks

The negative masks are obtained directly from the original using a panchromatic "tone" film and exposing through a colour filter.

The colour of the filter is the complementary of the colour of the mask: thus (Fig. 253) if we wish to obtain a negative mask of 40% of Cyan, which will serve to correct the Magenta printing, we must use the colour filter complementary to Cyan, that is, the Red filter.

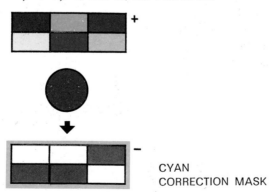

Fig. 253

CYAN
CORRECTION MASK

This Cyan correction mask will be used in register with the original, and from this combination, exposing through a Green filter, we proceed to the separation (already corrected) of the Magenta (Fig. 254).

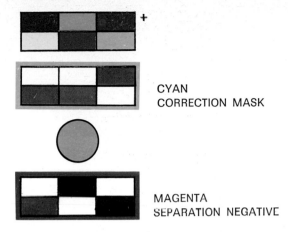

CYAN
CORRECTION MASK

MAGENTA
SEPARATION NEGATIVE

Fig. 254

The same procedure is repeated for correcting the Yellow separation with the 40% Magenta negative mask.

The same procedure can also be applied for making the 15% correction masks for the minor impurites.

From the corrected separation negatives we can then make the screened positives.

The *limitation of the method* lies in the impossibility of using the 100% pre-masks with a consequent loss of all the advantages explained in the "N. B." on page 170.

B) Method of correction with positive grey masks

The positive masks are obtained from the separation negatives using an orthochromatic "tone" film exposed with white light.

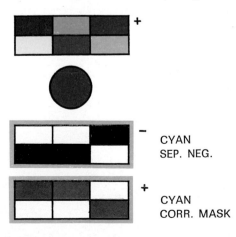

CYAN
SEP. NEG.

CYAN
CORR. MASK

Fig. 255

189

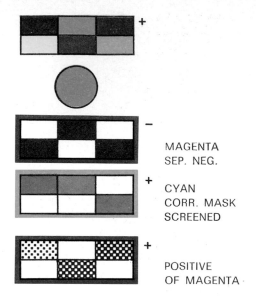

+

−

MAGENTA
SEP. NEG.

+

CYAN
CORR. MASK
SCREENED

+

POSITIVE
OF MAGENTA

Fig. 256

The Magenta mask is obtained from the Cyan separation negative (Fig. 255).

This Cyan correction mask is placed in register with the Magenta separation negative (Fig. 256) to obtain the screened positive.

The same procedure is repeated for correcting the Yellow separation negative with the 40% Magenta positive mask.

If desired 15% positive masks can also be prepared for the minor impurities.

The *limitation of this method* lies in the absence of 100% pre-masks with the consequent loss of all the advantages explained in the "N.B." on page 170. It is thus, as the reader may already have noted, a reduced version of the basic method described in the second section of this chapter.

C) **Method of correction with negative colour masks**

Films specially designed to obtain negative colour masks have been commercially available for some years (Tri-mask; Multimask). They are based on an opportune combination between the "spectral sensitivity" of the emulsion and the "colour of the colour coupler" of that emulsion.

It can be better explained in this way:

The colour masks are, like all colour films, composed of three or more sensitive layers. Each one of these sensitive layers contains a "colour coupler", namely, a chemical substance that in the development stage, confers to each layer a different colour (for example: a Cyan, a Magenta and a Yellow).

However, while in normal colour films the "colour coupler" of the single layers corresponds to their "spectral sensitivity" (that is, the Red sensitive layer of the original becomes Cyan, etc.), in these special masking films, the "colour coupler" (Cyan, Magenta or Yellow) of the individual layers corresponds to the spectral sensitivities of the impurity (major or minor) which it must correct.

Thus, for example, the layer containing the Magenta coupler, which will serve as the correction mask for the separation with the Green filter (Magenta separation) is sensitised only to Red radiations: after development, it gives a Magenta negative image with the maximum densities in the Red zones of the original, remaining transparent in those zones which contain Cyan (Cyan, Greens and Blues).

When this negative mask is combined with the original for making the separation with the Green Filter (the negative which will print the Magenta ink), the Magenta densities of the mask plus the Green filter (complementary of the Magenta) will give a grey more or less dense and will not make an impression on the separation negative, while the transparent zones (corresponding to the Cyan, Greens and Blues of the original) will blacken the separation negative.

This blackening (say of 40%) will reduce the amount of Magenta ink in the Cyan, Green and Blue zones, that is, those zones which are printed with Cyan ink.

The other colour layers work on the same basis to correct the other impurities.

Conclusion

To close our exposition of methods of photographic colour correction (sections 2 and 3 of this chapter) it is useful to remember that all these methods (fundamental and others) may be used with different techniques for making separations from both opaque originals and transparencies, and that the method we have called "fundamental" remains, not only from an instructive point of view, but also from a practical, unsurpassed and, perhaps, unsurpassable in the future.

SELECTION OF EXAMPLES

In the following pages we have collected a series of colour reproductions all carried out with the fundamental method.

The originals, from drawings and transparencies, have been chosen for their wide range of differences and subject matter.

Original transparencies are shown also in black-and-white to give their original sizes.

Fig. 257

Original (Fig. 257):
Ektachrome transparency
Rolleiflex with 80 mm Planar lens
Exposure 1/125 sec at f/8

Reproduction (Fig. 258) carried out with correction for only the major impurities ((only two 40% correction masks) according to the scheme illustrated in Fig. 228 and 237, without any manual retouching.

The photograph was exposed towards evening and almost against the light, and thus presents completely black zones (the clothes of the persons in the foreground and the cross). Note the exceptional sharpness of the detail, but above all the depth (three-dimensional effect) which arises from the separation of the various planes (cross, avenue of trees, church and mountains) that are characteristic of the fundamental method.

194

Fig. 258 - *Lourdes. View of the Basilica, at evening* (Photo D.M.A.).

Fig. 259

Original (Fig. 259)
Ektachrome transparency
Rolleiflex 6 x 6 cm with 75 mm f/3.5 Tessar lens
Exposure 1/125 sec at f/8.

Reproduction (Fig. 260) made with the correction of the major impurities with two 40% correction masks (as in the preceding example). In addition, with the object of giving more prominence to the yellow reflections of the vegetation, the Cyan separation negative was corrected with a 30% special effects mask as explained in Combination 1 on page 181.

Without manual retouching.

196

Fig. 260 - *Madagascar. Cultivation of cloves in the region of the Tamatale* (Photo Acquistapace).

Fig. 261

Original (Fig. 261)
Ektachrome transparency
Rolleiflex 6 x 6 cm with 80 mm f/3.5 lens

Reproduction (Fig. 262) carried out as the preceding example (Fig. 260), but with a mask of 20% (Combination 1).
Without manual retouching.

Fig. 262 - *Children with flowers* (Photo Farabola).

fig. 263

Original (Fig. 263)
Ektachrome transparency
Rolleiflex 6 x 6 cm
with 80 mm f/3.5 Planar lens
Exposure 1/60 sec. at f/8

Reproduction (Fig. 264) made by the same technique described for the example in Fig. 260, with the object of lightening the Cyan ink in the zones receiving the Yellow printing (leaves, house and field).

On pages 202 and 203 the "progressives" of the upper part of the picture are given with the object of showing the high quality of the separations which can be achieved with our method without any manual retouching.

A good working system can permit making up to 12 masked separations by contact from 6 x 6 cm transparencies (or 24 of 24 x 36 mm) in less than an hour, giving negatives ready for screening on an enlarger.

Fig. 264 - *Pessano, College Don Carlo Gnocchi. Autumn view of the park* (Photo Carla Morazzo).
On pages 202 and 203, Figs. 265, 266, 267 and 268.

Fig. 269

Original (Fig. 269)
Ferranicolor transparency
Rolleiflex 6 x 6 cm with 80 mm
f/3.5 Planar lens
Exposure 1/60 sec. at f/11

Reproduction (Fig. 270) made with correction for the "major" impurities (two correction masks of 40%).

The "progressives" in Cyan, Magenta, Yellow and Black are shown on pages 206 to 209.

Fig. 270 - *Bartres, near Lourdes. Doorway of the house where Bernadette Soubirous lived* (Photo D.M.A.).

Fig. 271

Fig. 272

Fig. 273

Fig. 274

Fig. 275

Original (275)
Ektachrome transparency
Rolleiflex 6 x 6 cm with 75 mm
f/3.5 Tessar lens.
Exposure 1/125 sec. at f/11.

Reproduction (Fig. 276) made with correction of major impurities only (two correction masks of 40%), without manual retouching. An enlargement of 3½ times.

Fig. 277 on page 212 shows an enlarged section of x 14 magnification.

Fig. 278 on page 213 shows an enlargement of the screen dots in the facing reproduction and is typical of a poster size reproduction.

Fig. 276 - *Greece. Macedonian landscape* (Photo Acquistapace).

Original (Fig. 279)
"Nativity". Tempera drawing
by Maestro Ernesto Bergagna,
School of Beato Angelico, Milan.
(45 x 35 cm.)

Fig. 279

Reproduction (Fig. 280) of the
upper part made with major 40%)
and minor (15%) impurities.

The "progressives" are given on
pages 215 and 216.

Fig. 280

APPENDIX

TABLES

This appendix includes the Chromatic Tables and others of practical interest. A table of logarithms is to be found on page 55 and one giving tangents on page 61.

COLOUR TABLES

The colour tables are of great value in judging the accuracy of a screened separation even before proofs are made.

They are printed with the three colours Yellow, Magenta and Cyan of the Kodak scale of colours (Footnote 1) on the basis of the following scheme (Fig. 285) in which the numbers indicate the dot percentage for each colour: 1 = 10%; 2 = 20%; 3 = 30% and so on.

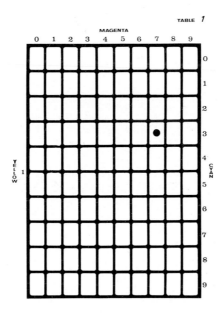

Fig. 285

The numbers on the left of the diagram indicate the percentage of Yellow, those above Magenta and those on the right Cyan.

For example, if we wish to know what colour corresponds to a screened separation which has a dot of 10% of Yellow, of 70% Magenta and of 30% Cyan, it is only necessary to look for the rectangle 173 on the table. It will be found in Table 1 (= Yellow 1, or 10% of Yellow) at the intersection of Magenta 7 (70% Magenta) with Cyan 3 (30% Cyan).

By comparing the colour of this rectangle 173 with the colour of the original, we can ascertain its accuracy and decide what retouching may be needed.

Footnote (1) The Kodak Scale is the most frequently used for three-colour printing with Yellow, Magenta and Cyan. It is also referred to as the « average scale » because it lies between the « DIN Scale » (cold) and the « European Scale » (warm). See note on page 128.

The Colour Tables given in the following pages number 12: of these 9 are in three colours (numbered from 1 to 9 in correspondence to the percentage of Yellow) and three in two colours: YM (Yellow and Magenta), YC (Yellow and Cyan), MC (Magenta and Cyan).

It should be noted that the tables faithfully reflect those which will result from printing, being themselves printed with impure inks.

These tables can also be used to arrive at a close enough approximation of the printing result from separations for gravure (positive in continuous tone), taking into account the relationship between the dot area and the density of the positive on the following basis:

Dot area	10%	20%	30%	40%	50%	60%	70%	80%	90%
Number	1	2	3	4	5	6	7	8	9
Density	0.30	0.45	0.60	0.75	0.90	1.05	1.20	1.35	1.50

The 100% dot area (solid printing) will correspond to a positive density for gravure of 1.65.

TABLE *1*

MAGENTA

0 1 2 3 4 5 6 7 8 9

YELLOW 1

CYAN

0
1
2
3
4
5
6
7
8
9

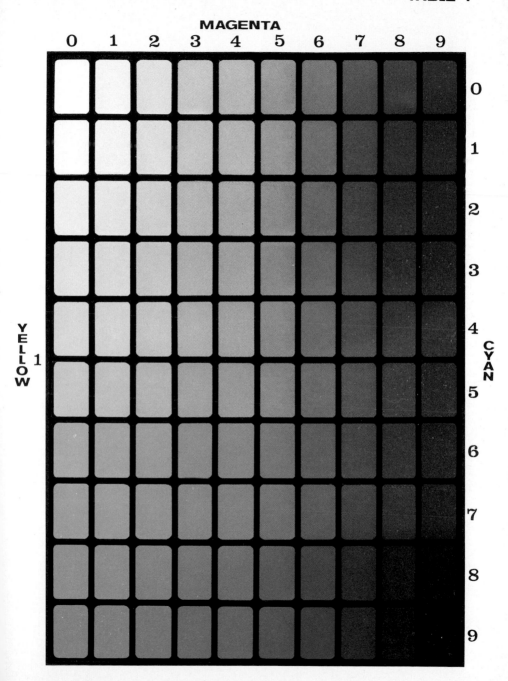

TABLE *2*

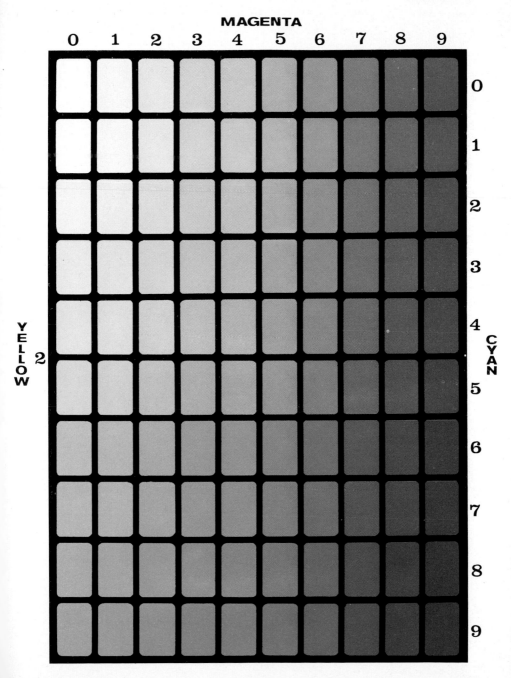

MAGENTA

0 1 2 3 4 5 6 7 8 9

YELLOW 2

CYAN

0 1 2 3 4 5 6 7 8 9

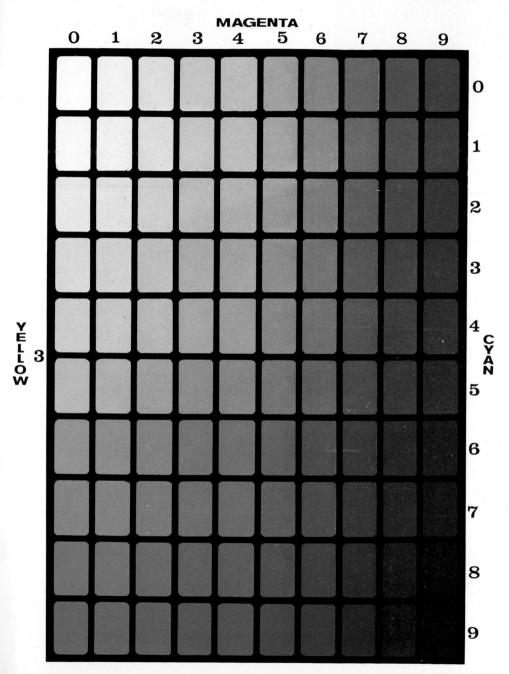

TABLE 3

MAGENTA

0 1 2 3 4 5 6 7 8 9

0
1
2
3
4
5
6
7
8
9

YELLOW 3

CYAN

TABLE 4

MAGENTA

0 1 2 3 4 5 6 7 8 9

YELLOW 4

CYAN

0 1 2 3 4 5 6 7 8 9

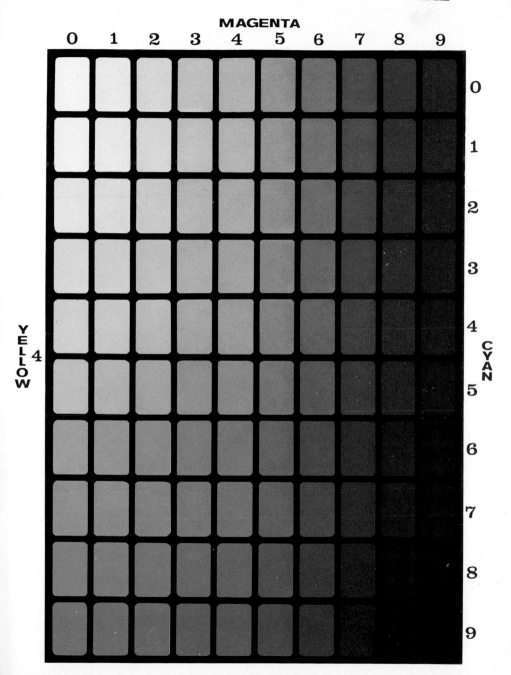

TABLE *5*

MAGENTA

0 1 2 3 4 5 6 7 8 9

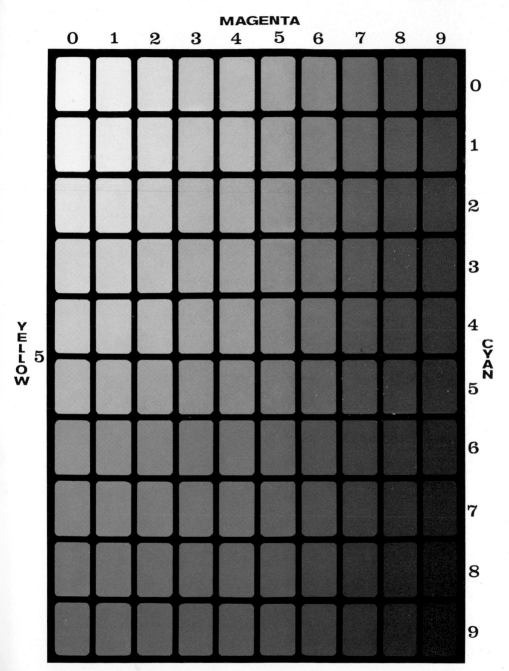

YELLOW 5

CYAN

0
1
2
3
4
5
6
7
8
9

TABLE *6*

MAGENTA

0 1 2 3 4 5 6 7 8 9

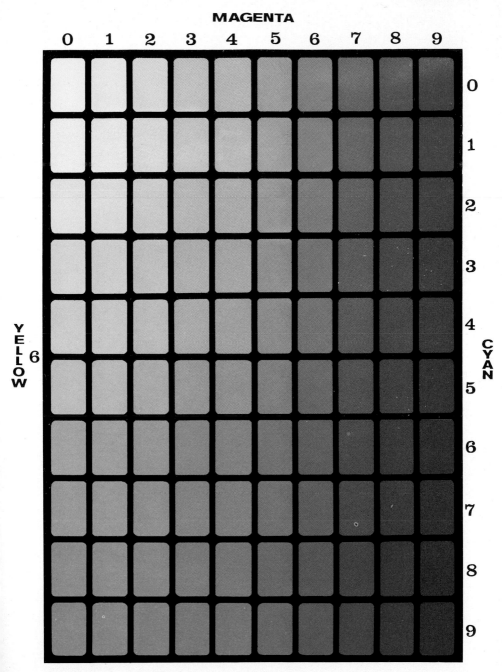

YELLOW 6

CYAN

0
1
2
3
4
5
6
7
8
9

TABLE 7

MAGENTA

0 1 2 3 4 5 6 7 8 9

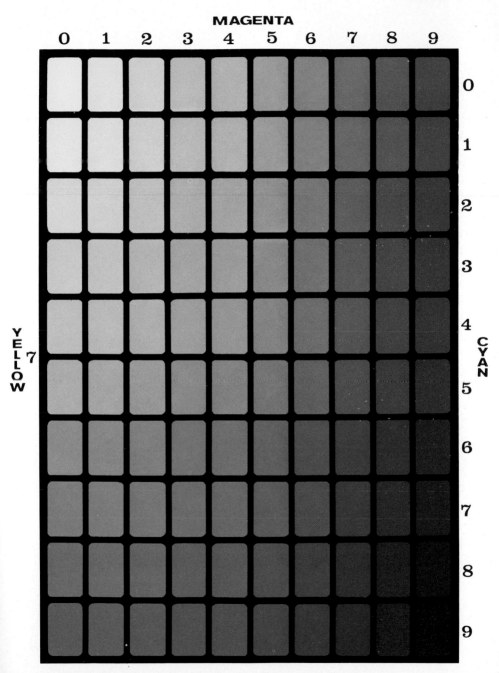

YELLOW 7

CYAN

0
1
2
3
4
5
6
7
8
9

TABLE *8*

MAGENTA

0 1 2 3 4 5 6 7 8 9

YELLOW 8

CYAN

0
1
2
3
4
5
6
7
8
9

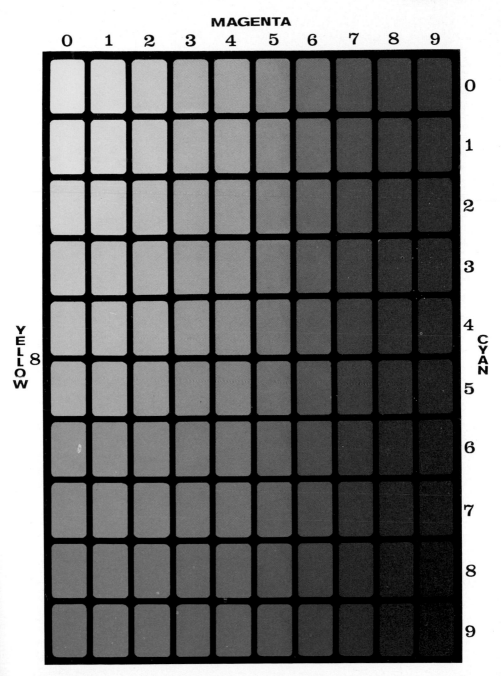

TABLE *9*

MAGENTA

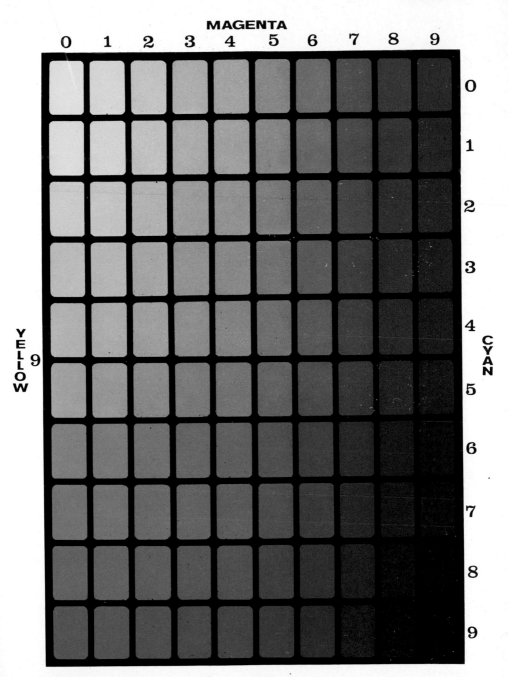

YELLOW 9

CYAN

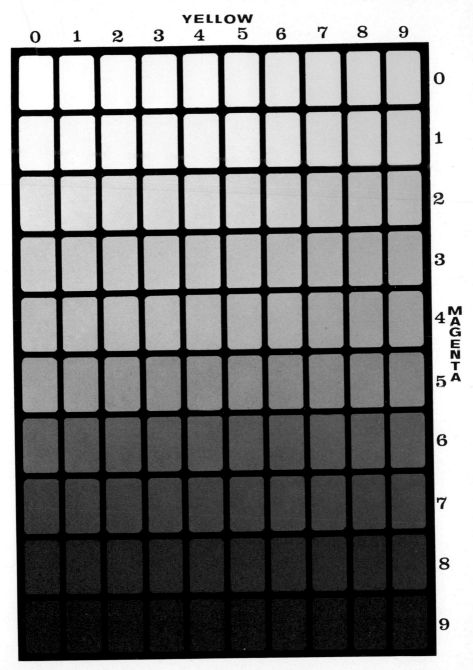

YELLOW

0 1 2 3 4 5 6 7 8 9

0
1
2
3
4
5
6
7
8
9

MAGENTA

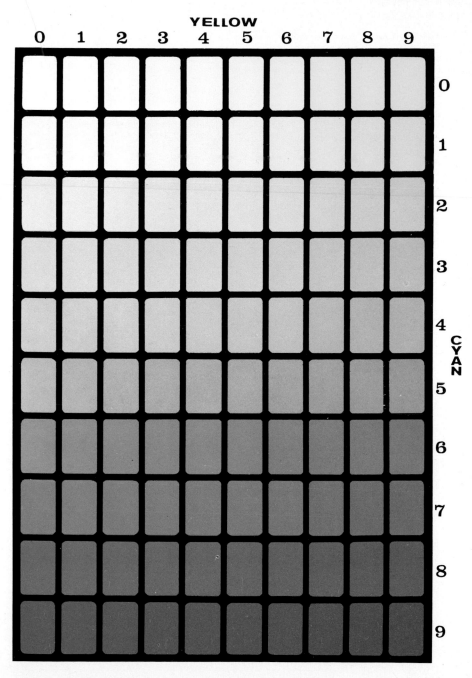

MAGENTA

0 1 2 3 4 5 6 7 8 9

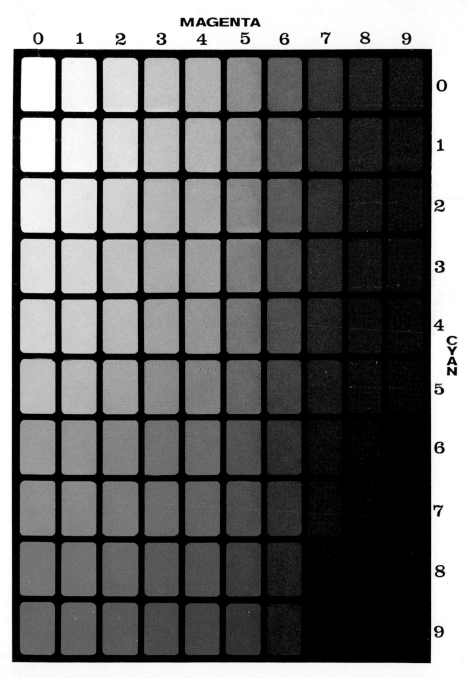

0
1
2
3
4
CYAN
5
6
7
8
9

ELEMENTS, SYMBOLS, ATOMIC NUMBERS AND WEIGHTS
(Oxygen = 16)

Element	Symbol	Atomic number	Atomic weight
Aluminium	Ac	13	26.98
Antimony	Sb	51	121.76
Argon	A	18	39.944
Arsenic	As	33	74.91
Barium	Ba	56	137.36
Bismuth	Bi	83	209.00
Boron	B	5	10.82
Bromine	Br	35	79.916
Cadmium	Cd	48	112.41
Caesium	Cs	55	132.91
Calcium	Ca	20	40.08
Carbon	C	6	12.010
Cerium	Ce	58	140.13
Chlorine	Cl	17	35.457
Chromium	Cr	24	52.01
Cobalt	Co	27	58.94
Copper	Cu	29	63.54
Dysprosium	Dy	66	162.42
Erbium	Er	68	167.2
Europium	Eu	63	152.0
Fluorine	F	9	19.0
Gadolinium	Gd	64	156.9
Gallium	Ga	31	69.72
Germanium	Ge	32	72.60
Gold	Au	79	197.2
Helium	He	2	4.003
Hydrogen	H	1	1.0080
Indium	In	49	114.76
Iodine	I	53	126.91
Iridium	Ir	77	193.1
Iron	Fe	26	55.85
Krypton	Kr	36	83.80
Lanthanum	La	57	138.92
Lead	Pb	82	207.21
Lithium	Li	3	6.940
Lutetium	Lu	71	174.99
Magnesium	Mg	12	24.32
Manganese	Mn	25	54.93
Mercury	Hg	80	200.61
Molybdenum	Mo	42	95.95
Neodymium	Nd	60	144.27
Neon	Ne	10	20.183
Nickel	Ni	28	58.69
Niobium	Nb	41	92.91
Nitrogen	N	7	14.008
Osmium	Os	76	190.2
Oxygen	O	8	16.00
Palladium	Pd	46	106.7
Phosphorus	P	15	30.975

Element	Symbol	Atomic number	Atomic weight
Platinum	Pt	78	195.23
Plutonium	Pu	94	242.00
Potassium	K	19	39.100
Praseodymium	Pr	59	140.92
Protactinium	Pa	91	231.00
Radium	Ra	88	226.05
Radon	Rn	86	222.00
Rhenium	Re	75	186.31
Rhodium	Rh	45	102.91
Rubidium	Rb	37	85.48
Ruthenium	Ru	44	101.7
Samarium	Sm	62	150.43
Scandium	Sc	21	44.96
Selenium	Se	34	78.96
Silicon	Si	14	28.09
Silver	Ag	47	107.880
Sodium	Na	11	22.997
Strontium	Sr	38	87.63
Sulphur	S	16	32.066
Tantalum	Ta	73	180.88
Tellurium	Te	52	127.61
Terbium	Tb	65	159.2
Thallium	Tl	81	204.39
Thorium	Th	90	232.12
Thulium	Tm	69	169.4
Tin	Sn	50	118.70
Titanium	Ti	22	47.90
Tungsten (Wolfram)	W	74	183.92
Xenon	Xe	54	131.3
Zinc	Zn	30	65.38
Zirconium	Zr	40	91.22

RELATIONSHIP BETWEEN PERCENTAGE TRANSMISSION AND DENSITY

% Tr.	D.	% Tr.	D.	% Tr.	D.	% Tr.	D.
0	3.00						
1	2.00	26	0.58	51	0.29	76	0.12
2	1.70	27	0.57	52	0.28	77	0.11
3	1.52	28	0.55	53	0.28	78	0.11
4	1.40	29	0.54	54	0.27	79	0.10
5	1.30	30	0.52	55	0.26	80	0.10
6	1.22	31	0.51	56	0.25	81	0.09
7	1.15	32	0.49	57	0.24	82	0.09
8	1.10	33	0.48	58	0.24	83	0.08
9	1.05	34	0.47	59	0.23	84	0.08
10	1.00	35	0.46	60	0.22	85	0.07
11	0.96	36	0.44	61	0.21	86	0.07
12	0.92	37	0.43	62	0.21	87	0.06
13	0.89	38	0.42	63	0.20	88	0.06
14	0.85	39	0.41	64	0.19	89	0.05
15	0.82	40	0.40	65	0.19	90	0.05
16	0.80	41	0.39	66	0.18	91	0.04
17	0.77	42	0.38	67	0.17	92	0.04
18	0.74	43	0.37	68	0.17	93	0.03
19	0.72	44	0.36	69	0.16	94	0.03
20	0.70	45	0.35	70	0.15	95	0.02
21	0.68	46	0.34	71	0.15	96	0.02
22	0.66	47	0.33	72	0.14	97	0.01
23	0.64	48	0.32	73	0.14	98	0.01
24	0.62	49	0.31	74	0.13	99	0.00
25	0.60	50	0.30	75	0.12	100	0.00

COEFFICIENTS OF EXPOSURE IN RESPECT OF DIFFERENCES OF ENLARGEMENT

Percentage enlargement	Exposure coefficient	Percentage enlargement	Exposure coefficient
10/100	0.30	260/100	3.21
20/100	0.35	270/100	3.40
25/100	0.39	280/100	3.60
30/100	0.42	290/100	3.80
35/100	0.45	300/100	4.00
40/100	0.485	310/100	4.20
45/100	0.52	320/100	4.40
50/100	0.556	330/100	4.60
55/100	0.60	340/100	4.80
60/100	0.64	350/100	5.00
65/100	0.68	360/100	5.25
70/100	0.72	370/100	5 50
75/100	0.77	380/100	5.75
80/100	0.81	390/100	6.00
85/100	0.86	400/100	6.25
90/100	0.91	410/100	6 50
100/100	**1.00**	425/100	7.00
110/100	1.10	450/100	7.60
120/100	1.20	475/100	8.20
130/100	1.32	500/100	9.00
140/100	1.43	525/100	9.70
150/100	1.55	550/100	10.50
160/100	1.68	600/100	12.20
170/100	1.82	650/100	14.00
180/100	1.95	700/100	16.00
190/100	2.10	750/100	18.00
200/100	2.25	800/100	20.00
210/100	2.40	850/100	25.00
220/100	2.55	900/100	25.00
230/100	2.70	950/100	27.50
240/100	2.86	1000/100	30.00
250/100	3.03	1050/100	33.00

COEFFICIENTS OF EXPOSURE IN RESPECT OF DIFFERENCES IN DENSITY

Density	Exposure coefficient	Density	Exposure coefficient
— 0.04	90/100	+ 0.04	1.1
— 0.08	83/100	+ 0.08	1.2
— 0.12	75/100	+ 0.12	1.3
— 0.15	70/100	+ 0.15	1.4
— 0.20	64/100	+ 0.20	1.58
— 0.25	56/100	+ 0.25	1.77
— 0.30	50/100	+ 0.30	2.00
— 0.35	45/100	+ 0.35	2.23
— 0.40	40/100	+ 0.40	2.51
— 0.45	35/100	+ 0.45	2.82
— 0.50	31/100	+ 0.50	3.16
— 0.60	25/100	+ 0.60	4.00
— 0.70	20/100	+ 0.70	5.00
— 0.80	15/100	+ 0.80	6.31
— 0.90	12/100	+ 0.90	8.00
— 1.00	10/100	+ 1.00	10.00
— 1.12	1/13	+ 1.12	13.00
— 1.20	1/16	+ 1.20	16.00
— 1.30	1/20	+ 1.30	20.00
ecc. ecc.			
— 2	1/100	+ 2	100

INDEX

absorption 129
acetic acid 35
acid compounds 23
Acigraph 98
additive synthesis of colour 121
Agfa Gevaert Multimask 190
Agfa Gevaert Verimask 167
agitation, development of halftone negatives 104
Alkaline compounds 23
ammonium chloride 36
angle acute 59
angle halftone screens 106
angle measurement of 58
angle obtuse 59
Angstrom units 119
anhydrides 24
anti-halo backing 28
apex of angle 58
arc lamps 138
arithmetic progression 58
artificial light 136
atoms 12-16
atoms size of 14
atomic linkage 21
atomic number 245-6
atomic particles 14
atomic weight 17, 245-6
autotype gravure 98
axis, X & Y on graphs 56

base compounds 24
basic chemistry 12
basic masking system 166
bi-metallic litho plates 99
black printing separation 148
blue-sensitive material 141
brightness of light 47
bromine 17, 19, 26
Bunsen & Roscoe, exposure 50

carbon 21
caesium 17
candela 49
cation 19
characteristic curve, construction 72-5
characteristic curve, interpretation of 72, 88

chemical development 33
chemical fog 33
chemical linkage 19
chemistry basic 12
chlorine 20
colormetric coordinates 131
colorimetry 117
colour 117
colour brightness of 128
colour correction examples 193-216
colour correction fundamental method 166-73
colour correction fundamental method examples 183-88
colour correction negative colour masks 190
colour correction negative grey masks 188
colour correction positive grey masks 189
colour correction pre-masks 170
colour correction principles of 161
colour filters 144
colour light 118
colour masking 161, 175, 178-9
colour masking highlight 176
colour objective 129
colour pigments 122
colour separation 143-4
colour separation deficiences of 148-51
colour separation diagram 146-7
colour separation examples of 173
colour specification of 128-135
colour specification C.I.E. 130
colour subjective 28
colour subtractive 123-8
colour tables or charts 219-43
colour temperature 139
complementary colours 121, 128
compounds 19
compounds classification of 22-5
contact screen 102
contact screen blue 105
contact screen classification of 106
contact screen contrast control 104
contact screen dot profile 102
contact screen magenta 104
contact screen yellow 105
contrast and density scale 71
contrast of reproduction 80

251

254